KEEPING PIGS

KEEPING PIGS

THE COMPLETE PRACTICAL GUIDE FOR PLEASURE OR PROFIT

Jeremy Hobson and Phil Rant

For Sarah – with love

David & Charles is an F+W Media, Inc. company
4700 East Galbraith Road
Cincinnati, OH 45236

First published in the UK in 2011

A catalogue record for this book is available from the British Library.

ISBN-13: 978-0-7153-3850-6 paperback
ISBN-10: 0-7153-3850-1 paperback

Printed in China by R R Donnelley
for David & Charles
Brunel House, Newton Abbot, Devon

Commissioning Editor: Neil Baber
Editor: Verity Muir
Project Editor: Ame Verso
Art Editor: Kevin Mansfield
Production Controller: Kelly Smith

David & Charles publishes high quality books on a wide range of subjects. For more great book ideas visit: www.rubooks.co.uk

CONTENTS

INTRODUCTION

'You dirty pig!' ...
'What a greedy pig!' ...
'You stupid pig!' ...
'You have the manners of a pig!'
These and many more such unkind epithets have given the much-maligned pig a very bad press. This book sets the record straight and gives this wonderful animal the long overdue recognition and respect that it deserves.

It is perhaps important to begin by acknowledging the enormous debt mankind owes to this humble creature. On a medical front, its body and organs have been utilized for many years to help improve our health. The pig's heart valves have been used to repair our own failing organs. The body of the pig – so remarkably similar to our own – has often been used to practise surgery techniques before they are performed on humans. The digestive tracts, also being similar in function and requirement to our own, have proved vital in testing new drugs.

Returning to the harsh comments with which we started; well, it may be that we use these slurs because we recognize the pig's similarities to ourselves! Pigs are not dirty by nature; they only become so if forced by circumstances of poor management and housing. Like all animals, pigs detest having to eat near to where they defecate and will always do so as far from their source of food as possible. The fact that they enjoy a cooling bath in mud on hot days is only to be admired, and is even emulated at luxury health spas for its beneficial effects.

As for greedy, have you ever noticed an overweight pig in the wild? It is unheard of. They will forage all day for nuts, roots and shoots in season, and when the bounty of autumn supplies a little extra they will eat a little more to store against the cold winter, but other than that they remain trim and fit. It is only when imprisoned by man in unsuitable housing without room for exercise or mental stimulation that they will eat too much and become fat, just like some humans!

Stupid? Most certainly not. Many people think horses are intelligent, but try putting a pig in a harness and getting it to pull a plough all day! No, they are too clever for that. As for riding pigs like horses, it has been done and with gentle training they will tolerate it for a while, but not for long. But try showing pigs just once how to operate a device that will give them food or a drink of fresh water and they will never forget, even after months of being away from that system. Given just a few minutes of investigation, a pig will soon learn how to open gate latches or to negotiate seemingly impossible escape routes. Cows and horses can often be seen standing in a grassless field looking over a makeshift baler-twine fence at a field of fresh clover, unable to work out how to overcome this simple obstacle, whereas pigs will soon devise an escape plan through or under all but the stoutest of enclosures.

The pig is a loveable rogue – a clean, inquisitive and ingenious animal capable of being a good companion to humans. Given a few simple requirements of housing, feeding and management, this noble beast will share your life with you and entertain you with its charm and active playfulness – look into that mischievous eye and it even appears to share your very thoughts. Pigs are almost always disgustingly healthy, annoyingly active and have very little in the way of food 'fads'. Few other animals can capture the heart and imagination in the same way as pigs. They can reward you many times over for the care you lavish on them and, if you so choose, will feed your family well when the time comes.

OPPOSITE With plenty of space and close enough to the house to keep an eye on the sow and her piglets, this is an ideal set-up.

ANATOMY OF A PIG

While veterinary knowledge is certainly not required in order to keep pigs, it is quite useful to know your hams from your hocks and your tusks from your trotters. The pig's body is streamlined (if you disagree, try keeping hold of one, especially if it's wet!) and evolved to enable it to pass easily through undergrowth. Depending on the breed, ears may be held erect or fall forwards and strong neck muscles allow for serious rooting about – not least of all under any less-than-sturdy fencing.

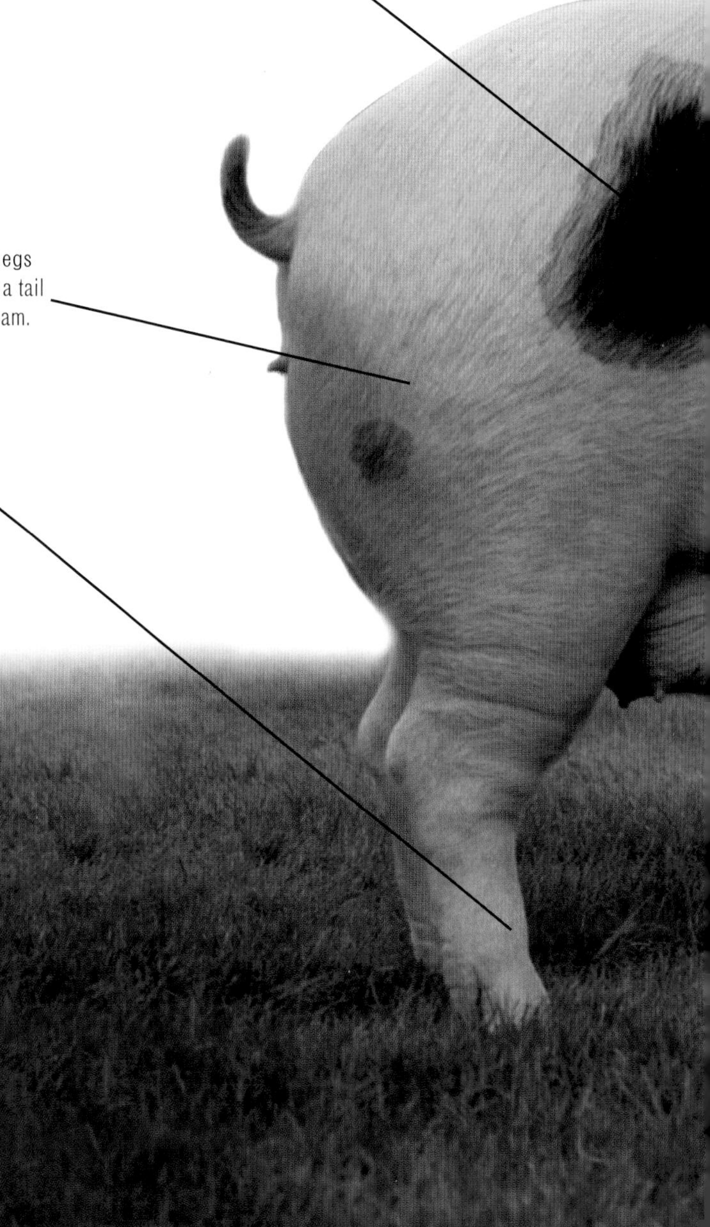

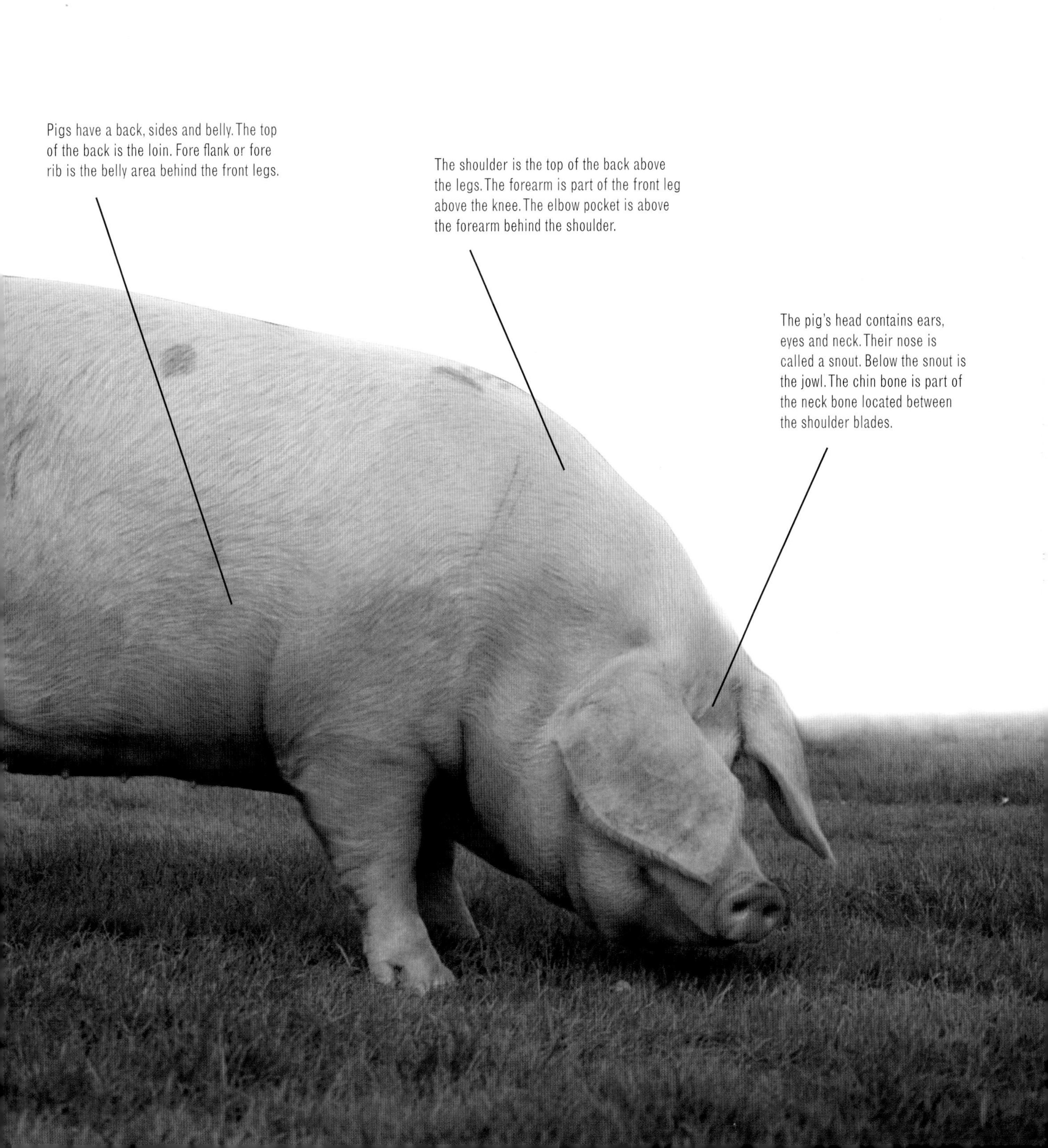
Pigs have a back, sides and belly. The top of the back is the loin. Fore flank or fore rib is the belly area behind the front legs.
The shoulder is the top of the back above the legs. The forearm is part of the front leg above the knee. The elbow pocket is above the forearm behind the shoulder.
The pig's head contains ears, eyes and neck. Their nose is called a snout. Below the snout is the jowl. The chin bone is part of the neck bone located between the shoulder blades.

PIGS IN ANCIENT HISTORY AND MYTHOLOGY

The word 'pork' originated in the French language and nowadays refers to meat while the name 'pig', which is actually Middle English, applies to the animal itself. The term 'swine' is Old English and 'hog' is quite often used as a generic term for a pig in America (although it may sometimes also describe a castrated male destined for the table).

The ancestors of the modern pig were semi-domesticated at about the same time as dogs – and certainly well ahead of other animals commonly seen in today's farmyard. Mesolithic tribes relied extensively on the pig as a domesticated animal some 2,000 years before they made the same use of sheep, cattle and horses. Outside the walls of Jericho in the Jordan valley, archaeologists have discovered remains that clearly indicate pigs were being butchered and eaten before 10,000BC: other animals only being domesticated around 2,000 years later as a result of Neolithic Man having lost his wanderlust and settling into a farming life rather than existing as nomadic hunters. In 7000BC, it would appear that human dwellings along the banks of the Swiss alpine river beds and lakes often included lean-to sheds, which must have looked very similar to the traditional cottager's pigsty – the reason for the lean-to system presumably being that the heat generated by its occupants would permeate through the house walls and help keep its residents warm.

TRADITION AND RELIGION

In Chinese astrology, every year belongs to an animal (with the cycle repeated every 12 years) and those who are born in that year are supposed to take on the animal's characteristics. Should you find yourself born in the year of the pig, Chinese horoscopes assess (or assassinate) your character as being, 'well-mannered, industrious, domesticated, but obstinate, egocentric and bad at planning.'

LORE AND LEGEND

The boar's head is a symbol of note in both legend and fact. A standard bearing a boar's head was among the gifts given to Beowulf by the Danish king in recognition of him having successfully fought and killed an ogre, while the soldiers of Valhalla were served a meal of boar's head stew from a cauldron named Eldhrimnir. The Saxons offered a boar's head as a sacrifice at the winter solstice – and perhaps the tradition of a roast pig's head with an apple stuffed in its mouth being brought to the table at Yuletide, is a continuation of this.

In the English village of Ripley, North Yorkshire, a boar's head appears on the coat of arms of the Ingilby family and commemorates an occurrence in 1357. Edward III was hunting wild boar in the locality when a large animal was flushed from the forest and presented the king with an opportunity to stick his spear into it. Unfortunately, the beast was only wounded and turned to charge the king's horse, which naturally panicked and threw his rider, leaving him in severe danger of being gored by the boar's large tusks. Luckily, Thomas Ingilby was there to witness the scene and he charged in to finish off the injured boar. As a reward, Edward knighted him and granted him the use of a boar's head emblem as his family crest.

In Hinduism, one of the prime incarnations of the god Vishnu was a male pig that went by the name of Varaha and it seems that a conflict between Varaha and Hiranyaksha (a demon who in other religions would have been the Devil) lasted for over 1,000 years before Varaha could take repossession of the Earth. Other faiths, however, did not hold pigs in such high regard, and ancient Buddhist thinking had it that the pig represented the 'sin' of desire, both materialistic and sexual.

In Ancient Egypt, pigs were commonly used for sacrificial purposes in the hope of ensuring good luck and fertility. So important were they that pig-keepers were esteemed members of the Egyptian community and are depicted on several tombs and sarcophagi, as well as in paintings and pottery. Interestingly, pigs experienced mixed fortunes in Egypt; by 1000BC, the animals were considered evil and those charged with looking after them were treated as the lowest of the low and were forbidden to enter temples.

In Ancient Greece, pigs would be lovingly reared before being killed and buried in sacrifice to Demeter, the goddess of fertility. In the subsequent spring, the pigs' remains would be disinterred and their remains combined with cereal seeds – the very definition of organic fertilizer!

ABOVE LEFT This 13th-century depiction of a pig being slaughtered shows how little the principles have changed.

OPPOSITE This detail of lions eating a wild boar is from a second-century Tunisian mosaic.

WHAT TO CONSIDER

PIGS ARE CHARACTERS – AND THAT IS REASON ENOUGH TO KEEP THEM. THEY ARE ALSO EXCELLENT ALLIES SHOULD YOU WISH TO CLEAR A ROUGH AREA, BUT THE ULTIMATE REASON FOR MANY MUST BE THAT THEY PROVIDE PERFECT PORK FOR RELATIVELY LITTLE OUTLAY. ASIDE FROM COMPLIANCE WITH ANY LEGISLATION ISSUES, ALL THEY REALLY REQUIRE IS COMPANIONSHIP, FOOD AND WATER, AN ARK AND SECURE ENCLOSURE, SHADE FROM THE SUN AND PERHAPS A WALLOW IN WHICH TO WHILE AWAY THE HOURS.

ABOVE In dry conditions, nuts spread on the ground rather than in a trough will help keep pigs entertained and busy.

OPPOSITE A sow and her piglets will soon become important members of your family.

SPACE

Space is the most important thing to consider when deciding to keep pigs: space in your daily schedule as well as physical space. Although you should always set aside a moment or two to talk to your pigs, general care and feeding need not take up much time on a routine morning and evening basis – and some of the chores such as cleaning could be a job for the weekend (but make sure that these are not neglected as one weekend has a nasty habit of turning into the next).

Physical space is another matter entirely and you need to work out exactly what breed of pig will best suit the space you have available. Because the Large White, for example, has been the mainstay of many a cottager's garden in the past, you might be tempted into thinking that a single pig needs no more than the traditional pigsty, which has a house and small concreted run attached, but there are flaws in the argument; not least of which is the fact that pigs should never be kept on their own as they are very sociable creatures. It is also important that pigs are given the opportunity to entertain themselves, whether it be by rooting about in woodland paddocks or simply having the opportunity to push an object around with their snouts – without such diversions, pigs, being very intelligent, will almost certainly become bored, which in turn can lead to habitual behaviour and a bad-tempered animal.

MOTIVES

Initially, you need to ask yourself why you want to keep pigs – is it as a hobby and to eventually breed and perpetuate one of the rare varieties, the offspring of which you can sell on, or is it to be able to put some meat in the freezer, in which case you will probably be starting with 'weaners' rather than adult breeding stock? If meat is your aim, the choice of breed is crucial as some types of pig are more suited to this than others. While you should never ignore a pig or fail to build up an empathic relationship, perhaps you should beware of getting too fond of an animal that is ultimately intended for slaughter. Your relationship with a breeding sow that quickly becomes part of the family is likely to be very different to one with a pig that you know is only with you for a limited amount of time: for that reason, consider how your children will feel (let alone yourself) if you decide to keep pigs simply for their meat value. In such a situation, perhaps it's best not even to give your pigs names.

NEIGHBOURS

If you live in close proximity to other people, it might be as well to pop round and tell them that you are considering a new hobby and to ask whether, in principle, they would have any objections. Most likely their first reaction will be to ask, 'But don't pigs smell?' to which you will be able to answer 'No, not really!' Pigs are the

PIG-KEEPING POINTERS

- Choose the right pig breed for your situation – there are some that produce outstanding pork, others that make ideal docile pets, perfect for your first venture into pig keeping, some that farrow easily and bring up their litter well and others that, through temperament and size, may be best left alone until you are more experienced with pigs.
- Pigs need shelter from both inclement weather and sunshine – remember that they can die of heat exhaustion.
- Don't keep a pig on its own. Conversely, don't overstock.
- Unless they are being kept totally for meat production, socialize with your pigs and get them used to human contact and your day-to-day routine.
- Depending on where you live, you might be able to obtain a grant if you are considering keeping pigs to clear woodland. You can often claim for funding to erect a perimeter fence and also to pay for the first planting of trees.
- In most countries, you will need to conform to various government edicts, possibly by registering as a pig-keeper, and also have a licence for whenever it is necessary to move pigs on and off your land.

cleanest of animals and, provided that you regularly collect their faeces (which they conveniently drop in more or less the same place each time), there should actually be very little to offend the noses of the neighbours.

COURSES AND COOPERATIVES

Before embarking on any form of pig keeping, it is wise to consider attending a basic pig husbandry course at your local agricultural college – there are plenty of opportunities on offer, some of which operate on a day or weekend basis. Details of up-coming courses are usually advertised in smallholding magazines or you can find relevant details on the college website.

You could also investigate becoming a shareholder in a city farm or local cooperative. Typically, how these things work is that all participants pay a percentage of forecast purchase and care costs and then, when the pig has been sold or butchered, receive a percentage of the profits. While it might not be satisfactory to all would-be pig-keepers, it nevertheless gives the opportunity for individual involvement where otherwise there might be none. Another advantage is that it is possible to care for pigs on a communal basis and learn a great deal about pig keeping while doing so – which will prove invaluable should the day ever come when it is possible to consider two or three pigs of your own.

ABOVE It is easy to see why pigs appeal to so many people – a litter of piglets snuffling around is simply irresistible.

SEMI-COMMERCIAL OR HOBBY?

The decision whether to keep fattening pigs on a semi-commercial basis or merely to enter into a small-scale hobby, say, with a breeding sow or two, must partly be decided by the space and accommodation you have available. These are truly important factors and will determine just what is possible to achieve.

Firstly, should you be considering breeding sows, you will need a piece of land – and it can be rough scrub land – divided into at least two paddocks (though three or more would be better), to keep them moving on to fresh land. Also, you will need somewhere for the second female to reside for a few weeks while the first is producing and raising a litter, so there should be two arks.

The cost of setting up your paddocks and arks for a small breeding enterprise could well add up; good stout fencing is required, or if you choose, an electric fencer, posts and wire, insulators and so on (remember that if electric fencing is used, someone should be at least available most of the day. A stray pig wandering your local area will, on a single occasion, most likely just cause some amusement, but if it becomes a regular occurrence you will quickly be reprimanded). Therefore, the practicalities and necessary finance will dictate the size and viability of your enterprise – never has the saying, 'you cannot make a silk purse out of a sow's ear' been so apt!

It might be that you have no thought at all of making money from your pigs and you just want to fulfil a yearning to keep a couple of pigs to show, or as family pets. When kept for the latter reason, it is important to realize that a pig can become very attached to its owner or indeed any other animal with which it lives. David Bland, who as well as being a poultry and pig authority of note, also runs a very successful feed merchants, tells of a customer who kept a pig and ewe sheep together; 'when the ewe eventually died, the pig starved itself to death. The vet could do nothing about it.'

ABOVE Pigs are sociable creatures and crave company whether it be of their own kind or the human variety.

PIGS AS PETS

Pigs that have been raised with constant human contact can make ideal children's pets although, of course, they are a little too big to cuddle! However, once the housing has been established and they have been taught how to make sure the door is properly secured, children will love to take care of a pig or two.

We would always advocate that there is an adult present: pigs are inquisitive creatures and their main method of exploring things is with their mouth – a small child may easily misinterpret this as an attempt to bite them and an adult close-by will reassure them, as well as ensuring that the pig does no harm. A child will also be fascinated by the sight of a sow with her new litter although this should always be viewed from behind the safety of a wall or partition, as a mother pig will be very defensive towards her newborn (as indeed will any animal). Cleaning out and feeding both hold a fascination for a child and to learn the source of the food on their plate is very important.

A good friend once related how his children had adopted a pair of pigs that he was rearing for the table. They had given the pigs names, spent a lot of time with them and took happily to the task of collecting suitable food from the hedgerows. He was surprised that they even enjoyed mucking out the animals. Naturally concerned as to their reaction when he had to kill the pigs, he had always told the children that this would be what was going to happen. The pigs grew – as pigs do – and eventually the time came to send them to slaughter.

Not long after the deed was done, the family had just sat down to a Sunday dinner that included a joint from one of the pigs, when the friend's youngest asked if they were eating his pig. Fearing tears, he said that it was not. On hearing this, the lad looked quite downcast and said, 'But Dad, I wanted a bit of *my* pig!', thus proving that children will never cease to surprise you and if taught the true facts of life, all will be well.

BELOW Pigs and children soon come to an understanding – especially if 'treats' are on offer.

HOW MUCH SPACE DO I NEED?

If you have a smallholding then you are halfway there, as you will have room for at least a few pigs. Unfortunately, not many of us are in that happy position, but it need not necessarily mean that you cannot rent some ground or, with just a little land around home at your disposal, tailor your plans to whatever is available.

All traditional breeds of pigs are basically outdoor animals and some are 'grazers' rather than 'rooters'. You should always aim to offer maximum rather than minimum space requirements, but the very minimum space for a pen of two adults would be a part-shaded area of about 300m^2 (around 3250sq ft), inside which would be a shelter or an ark. Better still is to have double that space and fence it in half so that one paddock can be used while the other is resting – of which more in chapter 4. Specific sizes for arks and the like are also covered in the same chapter, but it might be worth mentioning here that, if you are in the fortunate position of being able to house your pigs in a barn (with access to the outside), it will pay to make a part of it deliberately smaller by portioning a section off in order to create a cozy sleeping area in which your pigs will feel safe and secure.

Chapter 4 also deals with the subject of fencing, but it is nevertheless important to say here that some pig breeds, such as the Large Black are extremely docile and as such, this placid temperament enables them to be easily contained by two single strands of electric fence rather than having to construct a more secure post and stock-wire arrangement – which might be the case were you to choose another, far more independent breed. Having said that, if your pigpen is to be permanent, such fencing would be appropriate and give greater peace of mind – you most certainly don't need to be worrying about your pigs getting out and causing havoc in a neighbour's garden while you are out at work!

ABOVE Good solid stock fencing takes a lot of the worry out of pig keeping – and your pigs out of the neighbour's garden.

LICENCES AND LEGALITIES

Before considering erecting any new buildings or altering your existing ones, it makes sense to contact your local council and find out exactly what is and is not allowed under their particular by-laws and planning permission requirements.

Depending on the size of your venture, you may need to register with the relevant government agency. In America, it is best to check with the Department of Agriculture (USDA) about any possible zoning regulations for your land. In addition, some farm animals are not allowed to be kept within city limits. The non-government organization the American Farm Bureau Federation (AFBF), with offices throughout the US, should also be able to advise you on the current legislation.

In the UK, the government agency would be the Department for Environment, Food and Rural Affairs (Defra) and you need to obtain a County/Parish Holding (CPH) number for the land where the pigs will be kept. This is normally quite a simple procedure and merely involves contacting your Rural Development Service (RDS) who will take your details, give you your number and then send confirmation in writing.

MOVEMENT LICENCES

In the UK, pig movements usually take place under what is known as a 'general' licence – a copy of which can be downloaded from the Internet. Any livestock movements must also be recorded on a movement form (or even in a movement book if you are likely to be regularly transporting animals to and from your property). If you buy pigs from a market, an individual movement licence will be issued by the local authority trading standards officer so, if you think that you will eventually buy your pigs from such a source, it will pay to take a trip there well before buying any animals in order to find out more and acquaint yourself with the procedures. On a movement form will be recorded information such as ear-tag or slap-mark numbers, the date and time of movement, where the animals have come from, where they are going to and the CPH numbers of both places. The paperwork is multi-coloured, but all have a purpose! The yellow sheet is, for example, kept by the person from whom you have bought your pigs, the blue dealt with by the livestock transporter and the pink retained by you (which you must keep for at least six months).

ANIMAL TRANSPORT REGULATIONS

In Britain there are restrictions on transporting any animal further than a certain distance without first obtaining a special transport licence. It will also most likely be necessary that further movement to and from the property will be embargoed until the end of a 20-day standstill period. This is all designed to protect against the spread of possible

disease and the standstill is intended to act as an incubation period during which time any sickness should manifest itself. Exceptions may be made in the cases of pigs going to a show, for veterinary treatment or being serviced by a boar – all of which can be explained by your nearest Animal Health Office (AHO).

Once at home, new pigs may require to be registered with a local government authority – again you should contact your AHO. You will probably be required to secure a herd mark and/or number – even if you only ever intend keeping a couple of pigs. Pigs are then given a separate herd number (usually one or two letters appertaining to your holding, followed by four numbers), which are used whenever ear-tagging or slap-marking.

In America, ear-tags are used to facilitate animal traceability, and it is recommended that pigs should be isolated for a period of 30 or 60 days (depending on circumstances) after being transported to a farm or smallholding – again, check with the USDA or AFBF for further details.

WALKING LICENCES

Should you only ever be intending to keep a pet pig or two and want to take them from your premises – perhaps on a harness for a walk (yes, some people do!), you will need to obtain a walking licence from the AHO and abide by the following conditions: a) use the identical route declared on the certificate, b) always keep the pig(s) on a lead and, c) make sure they do not come into contact with other pigs. Your local officer may also set other conditions if he or she feels it necessary.

OPPOSITE Tags contain a wealth of information via the combination of letters and numbers and, in most countries, are a legal requirement.

ABOVE All pigs, regardless of breed, age or intended purpose, should be ear-tagged.

FACING REALITY

A couple of pigs as a hobby are fine, but should you be considering breeding from your pigs in order to ensure a continuation of stock, there may well come a time when you need to face the prospect of either selling off or sending to market those surplus to requirements. Of course, when buying a couple of weaners to put in the freezer eventually, the prospect is an inevitability and, while we have already established the suitability of pigs as pets, the reality for most of us is that pork is also very good to eat.

If you treat a pig with kindness, feed it well and keep it happy and content, provided the animal is dispatched in a humane fashion, as the law now dictates, then there should be no problems. A visit to your local abattoir beforehand will allay any fears you may have, and, when the time comes, you don't have to watch. The advantage of the process is, by arrangement, you can deliver your animal one day, and collect it the next, neatly butchered and packaged and bearing little resemblance to the friend you left the day before. Conversely, if you wish, you can collect the animal as two sides and some offal and do the butchering yourself (see chapter 8).

If you are raising animals for sale as breeding stock, or for someone else to fatten, sending them to market enables you to continue to rear more animals, perhaps to expand your enterprise, and while you will certainly not get very rich, the business should at least support itself. While briefly on the subject of killing for the table, the term 'fatten' is a bit of a misnomer although it is commonly used – the last thing that is wanted is a fat pig and a better term would be 'rear'.

At some stage, it will inevitably be necessary to have an injured, sick or aged animal put down. A veterinary surgeon once remarked to one of the authors, 'When you have livestock, you must expect dead stock' so, if that concept is unacceptable to you, it is perhaps best not to keep animals of any description.

BELOW 'This little pig went to market' – and it is probable that your surplus pigs will do too.

MALE, FEMALE OR BOTH?

Generally, and unless you are only ever intending keeping weaners to send to the abattoir, the question of the sexes depends entirely on what you intend to do with your pigs in the long run.

As hobby or pet pigs, there is no doubt that females will be your preferred option. From the point of view of rearing animals to be killed for meat, it will not matter, but you may need to consider the prospect of castrating male piglets to avoid the possibility of 'boar taint' – an unpleasant smell and taste emanating from the meat – which might occur in a male pig of six months or over (see also chapter 6). With breeds that reach killing weight quickly this is not necessary, but if you intend keeping your animal on a slower growth rate (traditional pig breeds come to weight over a longer period of time), castration might be contemplated. Should it be required, it is most commonly done by surgical means, as few other methods such as rubber bands or 'bloodless' castrators can be used on the pig due to the shape and position of the testes.

For those wishing to concentrate on breeding, keeping a male pig to service the females is a matter of economics. When you are keeping only two or three females for breeding, the cost of keeping a boar will be prohibitive. Artificial insemination (AI) is a possibility, but, if you have a local breeder near you who will be prepared to hire you the services of his boar, the cost of transporting your females to him will be far less than the cost of feeding a boar, having to keep him separate at times, and all the other inherent costs involved. Consider also that if and when you decide to keep female progeny back to enlarge your herd, or merely to replace a female, you will need a different boar from the gilt's father. The final deliberation concerns the fact that occasionally a boar can be a little tricky to manage: while rarely naturally aggressive, he may become very single minded at certain times of the year, particularly when he senses that the ladies require his attention.

ABOVE Keeping a boar is not always a practical option, especially for the hobbyist pig-keeper.

LIKELY COSTS

Set-up costs needn't necessarily work out too expensively, as much of the equipment needed for keeping pigs can be bought second-hand by perusing your local newspaper, the classified sections of smallholding magazines, the Internet, and land agent's details of farm sales in your area. Even if you are considering buying new, there are so many potential suppliers that they all need to keep their prices at a competitive level and so it will pay to shop around to see who is offering the best deals and packages.

Various suppliers can provide you with a 'pig starter kit' that comprises of an ark, an electric fencing energizer, 20 electric fencing stakes, 250m (270yd) of six-strand poly-wire, deep bowl mains drinker, galvanized feed trough, four steel hurdles and even a pig moving board (which, while not essential, is undoubtedly useful when moving pigs around – see chapter 7).

FENCING

Fencing is usually the biggest outlay: if you are intending to fence a paddock, orchard or wooded area permanently, it might well be worthwhile going to the expense of a post-and-rail fence in-filled with high-tensile stock netting, but otherwise, an electric fence will, 99.9 per cent of the time, keep pigs in their place – the area beyond the pig paddock must, however, also be secure in the event of any escapees. Incidentally, if using stock fencing, you might want to consider running a length of barbed wire at ground level underneath your perimeter fencing in order to deter digging – a prick on a sensitive snout might be all that is required for a pig to learn its lesson.

STOCK

The cost of the pigs themselves will vary and will depend largely on whether they are pedigree, pure-bred, cross-bred or hybrid (see chapter 2). The age at which you are buying your pigs will also be reflected in the price – younger stock will generally be less expensive but then, if you are buying them for breeding purposes, you should consider the extra feed needed to rear them until they are mature enough to do so.

FOOD

Free-range or outdoor pigs can, if given enough space, obtain a proportion of their food by foraging but even so, will require some supplementary feeding, usually in the form of pelleted concentrates, commonly called pig nuts. Exactly how much food your pigs will need on a daily basis will, of course, depend on their breed as well as their size – youngsters obviously not requiring as much as adult animals, and non-breeding pets less than a lactating sow who is feeding her piglets. If at this stage, you have an idea as to what breed of pig you'd eventually like to keep, then perhaps it might be possible to gain some idea as to the likely cost of foodstuffs either by seeking advice from an experienced pig-keeper, or from your friendly local agricultural suppliers and feed merchants.

ADDITIONAL COSTS

Although it is unlikely to be crucial at the outset, you might, if the pig-keeping bug strikes really hard, need to consider the purchase of a pig trailer – especially if you decide to take up showing on a regular basis, or even simply in order to transport your sows to the boar on the occasions when they need serving. Small, specially designed pig trailers (see chapter 3) are relatively light and can be easily towed behind the normal family car but if your vehicle was bought in the interest of leaving the least possible carbon footprint, it might not be capable of pulling the trailer when it is loaded with an adult pig. Not that we are suggesting adding the expense of a new car to the list!

Don't forget incidental costs, which might include something as minor as a few ear-tags and annual membership of your local smallholder's club, or will be far more major should you need to call out a vet for a sick animal. There will also be the possibility of needing to purchase over-the-counter medication such as wormers and similar. Finally, you might like to consider a subscription to a smallholder's or pig-keeper's magazine – as a newcomer to the hobby, you will undoubtedly find much to educate and interest between its covers.

BELOW With enough space, free-range or outdoor pigs can obtain much of their food by foraging.

BREEDS

LARGE, SMALL, A 'ROOTER' OR A 'GRAZER', HAIRY OR RELATIVELY SMOOTH-SKINNED, SELF- OR MULTI-COLOURED, PURE-BRED OR HYBRID – JUST SOME OF THE MANY OPTIONS OPEN TO YOU WHEN YOU EMBARK UPON THE INTERESTING AND FASCINATING WORLD OF PIG KEEPING. SOME BREEDS WILL SUIT THE INDIVIDUAL BETTER THAN OTHERS, BUT ALL WILL SUIT SOMEONE SOMEWHERE! TIME SPENT IN CHOOSING THE RIGHT BREED FOR YOU IS NEVER WASTED AND CAN GO A LONG WAY TOWARDS AVOIDING POTENTIALLY DISASTROUS MISTAKES.

CHOOSING THE RIGHT BREED

Choose your breed carefully. The expression 'horses for courses' is well known but there should perhaps be another along the lines of 'pigs for purpose'. Some breeds lend themselves to an indolent life and are quite content to wander around their paddock or an orchard grazing and eating whatever is available rather than acting like mini-bulldozers. On the other hand, you may wish to keep a breed that will clear overgrown woodland of scrub and bracken; with its hardy nature, relatively long legs and strong jaws perfect for root and stem destruction, a small herd of Tamworths is the perfect solution in this case.

Consider your own strengths and weaknesses too. If you're not physically robust enough to put up with the inevitable exuberant bashing of the back of the legs or an uneven contest of snout-against-arm wrestling as you deliver dinner in a bucket, it will – all other requirements having been taken into account – pay to choose one of the smaller, quieter breeds.

ABOVE Large Whites were first recognized as a pure-breed in 1868.
BELOW LEFT A Micro-Pig is most certainly not a pure-breed.

PEDIGREES AND PURE-BREEDS

It is important to realize that there is a subtle distinction between pedigree stock and pure-breed stock. For the former, the breeder must be a member of a national association that acts as the governing body and maintains the relevant herd books. Any pig offered for sale should be registered in some way (normally by ear-marking), have been certified at birth and eventually registered with the association. A pedigree pig must come supplied with 'papers', all of which should be filled out and ready to collect at the same time as your pig. It is doubtful whether such a scenario would ever happen, but if the pedigree is not ready, do not let the breeder fob you off with a promise that they will 'put it in the post' – make it clear that it's a case of 'no pedigree, no purchase'.

On the other hand, pure-breed stock are not likely to have much, if any, in the way of documentation: they will, in all likelihood, be less expensive, but there will be no pedigree as such and so it will be a question of having to rely on the seller's honesty regarding an animal's parentage and its breeding. Where unrelated pure-breeds are in short supply, it may be that in order to keep certain bloodlines and necessary genes, breeders have had to resort to inbreeding and while this is a method often used to 'fix' desirable genes, without a complete knowledge and understanding of the procedure, undesirable genes may also be propagated. Forced either into accepting such a possible situation or going without, it will pay to source your initial stock as carefully as possible and to try and gain some reassurance from the breeder that their animals have not been subject to inbreeding in recent generations.

ONLY FOOLS RUSH IN'

Because of the need for suitable transport and the very real possibility that a movement licence will be necessary, it is unlikely that you will succumb to a whim and buy pigs on impulse as a result of visiting a show or auction. This scenario is often seen at a poultry auction, where inexperienced buyers, seduced by '*The Good Life*' factor and the prospect of a regular supply of fresh eggs, buy birds only to realize later that they have no suitable food or housing. Unlike chickens, pigs cannot easily be looked after by a neighbour when you are away on holiday, unless of course, they are as interested in your hobby as you are.

It is said countless times in this book (and the authors make no apology for doing so), but do choose your breed of pig wisely. Visit as many agricultural shows as you can, speak to all who are interested and read all there is to read before making your final decision. If you make the wrong choice you will probably give up and the pig world will have lost a much-needed convert. Make the right choice and it will gain a life-long enthusiast who, most importantly, will encourage future generations.

BELOW Despite being one of the oldest pig breeds, Tamworths are listed by rare breeds associations as 'threatened' in the US and 'vulnerable' in the UK.

BENTHEIM BLACK PIED

This extremely rare breed is well worth noting because of its German ancestry and also the UK's Berkshire bloodlines that run through examples of today's Bentheim Black Pied. Included on the breed list of American pigs, you would however, have to travel far and wide before seeing an example. Like many others throughout the world, the breed nearly became extinct in the late 1950s due to the somewhat narrow-minded activities of commercial pig producers who were intent on keeping and breeding large white-fleshed pigs that would give birth to big litters – which would in turn mature for the meat market as quickly as possible.

Known as the Buntes Bentheimer Schwein in its native Germany, the title incorporates the name of a region in Lower Saxony where it first appeared at the turn of the last century as a result of pig-breeders using imported English-bred Berkshire boars to mate with examples of their native farmyard pig. The resultant animals were further refined until the breed bred true to type and a herd book could be created. Sadly, subsequent commercialism all but killed the Bentheim Black Pied and use of the herd book ceased in 1964. A handful of dedicated German breeders did, however, continue to keep the few existing bloodlines alive and further safeguarded them by ensuring that small breeding herds were distributed throughout the country in order to lessen the risk of local disease killing off what stocks remained. Eventually a new herd book was created and the Association for the Conservation of the Bentheim Black Pied Pig, set up in 2003, brought things right up to date by using computers to regulate and assess the compatibility of male and female bloodlines – a pig dating agency, if you like! With such a small nucleus at their disposal, it was obvious that some inbreeding would be necessary and it was, therefore, extremely important that it was sensitively done. Efforts seem to be paying off and, not only have the numbers of breeding pigs risen, so too has the interest in the breed in both Germany and abroad.

Despite their name, which might lead you to assume they are predominantly black in colour, they are actually white with black 'pied' markings, which are themselves surrounded by a soft grey outer marking. Their ears fall forwards rather than stand pricked, and they are considered to be medium-sized animals.

RIGHT A new generation of Bentheim Black Pied.

BERKSHIRE

It is perhaps appropriate that the Berkshire follows on alphabetically from the Bentheim Black Pied (see previous entry), as exports from England of the former were instrumental in the creation of the latter. An old breed (indeed it is claimed by some that it is Britain's oldest 'true' breed), popular legend has it that they came into prominence nationwide in the 1600s after being taken from their native county by Oliver Cromwell's troops during the English Civil War. Realistically, however, records show that the Berkshire breed as it is seen today came into existence only in the early 1800s. The breed's rapid growth in the ensuing years is credited with providing a need for the first ever UK pig society being formed in 1883 and the first edition of the British Berkshire Society's herd book being published in 1885.

Eventually exported to Europe, America and beyond (in Japan, the meat of the Berkshire is highly prized as *kurobuta* or 'black pork'), the breed may be doing well generally but its fortunes do not look so promising in New Zealand where it is estimated that there are now fewer than 100 pure-bred breeding sows. Undoubtedly a very pretty-looking pig, it is, however, more likely that the Berkshire has survived the centuries because of its ability to do well in confined areas (making it an ideal cottager's pig) and the boar's siring ability. When all these factors are combined they form a very tempting package for the hobbyist who is seeking looks with an ease of keeping and a breed tolerant of inexperience.

According to breed standard recommendations, the Berkshire must be black overall and should show six white points: the tip of its tail, the length of its snout and at the base of each leg. Furthermore, the snout should be concave and, while the ears are carried erect, ideally they ought to be tilted slightly forwards.

RIGHT The Berkshire is a medium-sized breed of very neat appearance.

BRITISH LOP

No, not a type of rabbit as you might think, but actually a pig breed! Along with the Middle White (see separate entry), it is rated as being 'endangered' by the Rare Breeds Survival Trust (RBST) because there are thought to be 200 or fewer registered pure-bred breeding sows (the Middle White faring slightly better at around twice that number). The Lop can, therefore, perhaps be compared with the American Mulefoot (see separate entry) in its rarity. When the first herd books were created, it went under the gloriously convoluted name of the National Long White Lop-Eared.

Originating in the West Country, it is an all-white, obviously lop-eared pig that is extremely hardy, and has always been considered as equally well suited to commercial outdoor pig units as it has to supplying the needs of the smallholder. They can be kept outdoors, with the shelter of an ark for extreme weather and when farrowing. The sows are known to take readily to the boar, produce good-sized litters and make perfect mothers, often rearing their entire litter of piglets to maturity.

For those interested only in its capabilities as a meat producer, it provides a well-muscled, lean carcass and is sometimes used for crossing with commercial pigs to create a hybrid; while for those wanting a friendly, docile and easily managed pig it is one of the more obvious choices for the first-timer.

BELOW British Lops are extremely hardy and will work tirelessly for their supper.

DUROC

Whereas many pig breeds found their way from Britain to America (a logical direction when you consider that settlers would have taken with them livestock well established for several generations), the Duroc is an import from America to the 'old country'. Originating as a breed through the efforts and dedication of Isaac Frink of Saratoga, New York, in the early 1800s, and really into their stride by the late 1800s, it was not until the 1970s that the Duroc was seen in the UK.

Among its attributes is the fact that, according to the British Pig Association (BPA), 'their thick auburn winter coat and hard skin allows them to survive the cold and wet of the British winter. This coat moults out in summer ... as a consequence it can cope with hot dry summers equally well.'

Noted for their fast growth and excellent feed efficiency, they are well worth considering when costs need to be kept down and yet a meat carcass is required in the end. Although a white Duroc is known, it is in fact a cross-breed and the colour of all pure-bred animals should be 'red' – ranging from a yellowish-gold to reddish-black. Their ears are carried partially drooping.

They are reasonably easy to source, especially in the US where they are the second most recorded breed of pig in the country. In Britain, their numbers are flourishing, while elsewhere, such as in New Zealand, they are popular as a sire when it comes to cross-breeding for commercial purposes.

ABOVE Commercially, Durocs are often crossed with Large Blacks, Whites and Landrace – as a pure-breed, they are prolific breeders.

GLOUCESTER OLD SPOTS

The Gloucester Old Spots is a large lop-eared breed and is a good all-rounder as it provides tender, succulent, fine-grained meat that is particularly suitable for pork production (and especially bacon), due to the significant depth of its body. It is extremely hardy and, in most parts of the world, can be kept outdoors all year round with the inclusion of shelter and an ark for farrowing.

Although a rare breed – currently classed as 'critical' by the American Livestock Breeds Conservancy (ALBC) and registered at 'category 5, minority' by the RBST in the UK – it is by far one of the most popular of rare breeds and even those with only a modicum of pig knowledge will be able to recognize it. Unlike several traditional breeds who can trace their ancestry quite clearly, one can only make assumptions as to the actual bloodline origins of the Gloucester Old Spots: it is likely that it is an amalgam of several breeds that were periodically mated with a pig type known in the Berkeley Valley of England.

They are very docile and easily managed – many breeders have it that all lop-eared breeds are easy to manage because their ears prevent them from seeing too much and therefore becoming stressed and panicked, but make of that what you will. Gloucester Old Spots sows make excellent mothers, lactating easily and thereby providing a plentiful supply of milk. They rear good-sized litters long after those of other breeds have reached the end of their breeding life. Unusually, the boars are said seldom to pose a threat to piglets. All in all, it is possibly one of the best options for the amateur pig-keeper whether they simply want a human-friendly pet that will give them much pleasure, or a meat-producing animal for semi-commercial reasons – a perfect combination.

Correctly, when either writing or speaking of the breed, it should always have an 's' at the end of the name – even if you are speaking of a single animal. Predominantly white with black spots (in recent years, selection has been towards even less black), it is sometimes known as the Orchard Pig due to the fact that it was often found foraging in the apple orchards of its native English county. Somewhat less attractively, it was at one time occasionally referred to as the Swill Pig. In 2010, the European Union Commission awarded the breed Traditional Speciality Guaranteed status (TSG being one of the protected food names registered in Europe).

RIGHT A contented Old Spots in ideal conditions.

HAMPSHIRE

Rural residents of the county of Hampshire are still sometimes known as Hampshire Hogs! The term is not in any way meant to be disrespectful; more to do with a time when the breed was extremely well known in this part of southern England and most country-dwellers kept a pig of this type in their backyard. In fact, it appears that the term has been in use for several centuries – at least according to Francis Grose, who, in his book, *A Provincial Glossary*, published in 1790, described it as being a 'jocular appellation for a Hampshire man; Hampshire being famous for a fine breed of hogs, and the excellence of the bacon made there.'

Somewhat bizarrely, given the fact that it is named after a central southern county, it seems that this particular pig was at one time just as popular in the northern parts of England and also over the border into Scotland. This is perhaps explained by the fact that the breed is known to be extremely hardy and for its ability to thrive by foraging on almost nothing – and therefore could be expected to do well on the poorer soils and wilder topography. It was, however, from its home county that the first examples of the breed were exported to America during the early 19th century where it was originally known by a variety of names – including the McKay, McGee and Ring Middle.

Hampshire pigs are black with a white belt that circles the body at its forelegs. The ears are pricked and, according to some breed descriptions, 'large, erect and open.' It is purposeful and proud in its carriage and movements.

BELOW The long back of the Hampshire makes it an ideal bacon producer.

KUNE KUNE

Whether the name should be split into two separate words or connected with a hyphen is a moot point – in some references it is, in others it isn't. Whatever is decided, it is pronounced 'coonie coonie'. Thought to have originated in New Zealand where the Maoris used it as a commodity with which to barter and trade, the breed type is in fact more likely to have been taken there from Asia by foreign traders and explorers.

Like the Vietnamese Pot-Bellied (see separate entry), it has gained something of a cult following as a pet which, given their extraordinary liking of human contact, is perhaps not surprising. Devotees of the Kune Kune will tell you that they are better pets than the Vietnamese Pot-Bellied as they are far more easy going and less inclined to become snappy – thus making them interesting children's pets.

Unlike many other breeds, the colours are many and varied, and include self-colours such as black, brown, white, tan, gold, ginger and buff. In addition, there are also specimens that include pied markings, incorporating ginger and black, and black and white. Not content with coats of many colours, the hair type varies too: it ranges from fine through to coarse and curly. One thing that doesn't change are the characteristic tassels that hang from the animal's lower jaw, although not all individuals will have them and some may only have one. These tassels are properly called piri piri (or pire pire) and may be long and thin, or short and wide. Their ears can be either erect or semi-lopped. Being primarily grazing pigs, the Kune Kune should have access to good quality hay at times when grass is not readily available.

ABOVE The Kune Kune – snub-nosed, short-legged and full of character.

LANDRACE

There are several types of Landrace, many of which are prefixed by the name of a country. There is, for example, the American Landrace, which, unsurprisingly, is well known in America where it is the fourth most recorded breed. Confusingly though, there are also listed in America, the Belgian, British, Danish, Dutch, Finnish, French, German, Italian, Norwegian and Swedish Landrace. All types are, however, thought to have originated from Sweden as a white, lop-eared animal. All are also medium to large in size and have long snouts.

No matter what country the prefix might mention, the Landrace breed has been long loved for its docile nature, quiet temperament and for being good with its piglets. It has always been popular among pig-breeders for its long, lean profile and excellent bacon, although there are some that even now maintain certain strains possess weak backs and an inability to adapt in areas where a particularly hardy pig is needed. While that might well have been the case when the breed was bred purely as a commercial animal, hobby breeders and purists have almost completely counteracted the supposed problem since it was first noticed in the 1970s.

As if one problem wasn't enough, at one time the Danes refused to allow the exportation of the country's own strain of Landrace, fearing that to do so might result in the breed losing some of its characteristics and production quality if mated with inferior foreign strains. More likely, however, was the realization that another country might usurp Denmark's position as the premier producer of bacon! Today, the progeny of their exported animals can be found virtually worldwide.

In Britain, the breed (already known in post-war years) became very popular as a result of the Howitt Report of 1955 – a report that very nearly signalled the end of 'minor' meat-producing breeds due to its insistence that, in order to supply the industry efficiently, only the Landrace, Large White and Welsh should be considered, not least because of their undoubted ability to perform well under either indoor or outdoor systems.

The Landrace is, primarily, a meat producer and, arguably, lacks too much in character to be the perfect choice for the hobbyist (although it may have merit for the table-orientated smallholder). Having said that, it is perhaps worth including the comments of the BPA when it says that, 'The greatest strength of the Landrace is its undisputed ability to improve other breeds of pig when crossed to produce hybrid gilts … for the profitable production of quality pig meat.'

RIGHT A Landrace 'as happy as a pig in mud!'

LARGE BLACK

Large Blacks are one of the oldest breeds of pig and possibly one of the most characterful. Like the Gloucester Old Spots, they are lop-eared, extremely docile, hardy and well suited to simple outdoor systems – indeed many breeders insist that the Large Black can be kept safely contained by the use of only a single strand of electric fencing. The sows are excellent mothers with exceptional milking ability and are able to rear sizable litters on unsophisticated rations. Their black skin helps protect them from sunstroke and although the skin is black, the meat is not.

Research suggests that the Large Black actually originates from the Old English Hog (an ancient semi-wild type that scavenged for its food in places such as the New Forest), although elsewhere is mentioned the fact that they may well have evolved through the bloodlines of Asian pigs brought into England by sea-traders. Whichever, there were two very different types well established in both East Anglia and Cornwall by the latter half of the 19th century. Numbers grew and by the early 20th century, the Large Black could be found virtually everywhere in the UK, although it was not until the mid-1930s that the breed began to be exported overseas.

Sadly, whereas in 1921, the herd book apparently held over 10,000 registrations, the RBST calculates that nationwide in the UK today there are fewer than 400 breeding sows. Although the situation may be marginally better in America, the ALBC nevertheless has concerns for the breed's future – a good enough reason alone to consider the Large Black.

BELOW Renowned for its succulent meat, the Large Black is outstanding as pork but excels when traditionally cured as bacon.

LARGE WHITE

Should you ever have seen old paintings or photographs of a pig standing on its back legs in order to see better over the door of its sty, then it's a pretty safe bet that it is a Large White. They are known for taking a great interest in their surroundings and for watching what is going on around them.

A prolific breed, usually producing big litters of quick-growing piglets, it converts food efficiently – a factor you may find important should you wish to sell their meat or put a carcass in the freezer. On the down side, they are extremely large pigs and should you be contemplating a boar to run along with females, it must be noted that they can grow so heavy that the sows are unable to take his weight when mating.

First recognized as a pure-breed in 1868, a British herd book was eventually published in 1884 – the bloodlines from which suggest that, like the Middle White and Small White, it is a result of selective breeding between several pig types already known in Cumberland, Leicestershire and Yorkshire. In the US, it is known either as the American Yorkshire or the English Large White.

Today, the Large White is widely recognized in many countries because of it having been exported from Britain in order to create a commercially viable out-cross with existing native breeds. Despite its past popularity, in the days when almost every rural dweller had a pigsty, its size does not make it the best choice for those with limited space. On the other hand, as a meat producer (either pure-bred or when used to cross and improve other breeds), it may have some merit.

ABOVE The Large White is a rugged and hardy breed that can withstand variations in climate and other environmental factors.

MANGALITZA

Without a doubt the most unusual of all the pig breeds, the Mangalitza (or Mangalitsa) has a woolly, sheep-like coat and there have been many people unsure as to whether they are looking at a pig or a sheep! Because of its relatively recent appearance in the west, it has attracted much media interest and consequently, like the Micro-Pig (see separate entry), has a bit of a cult following, which has resulted in ridiculous prices being demanded by some breeders. Should you be seriously interested, then it might be better to wait until the furore has died down – as was the case when the media attached itself to Kune Kune and Vietnamese Pot-Bellied pigs, the price will drop once the hyperbole lessens.

The breed is native to the areas of former Austria-Hungary, but it is thought that they have a genetic link to the Lincolnshire Curly-Coat (supposedly the only breed ever to be sheared on a regular basis – its wool being used to make sweaters), which became extinct in 1972. Although still unusual in the UK (where the foundation breed was actually from Austria) and the US, in their main country of Hungary (where they are called 'Mangalica'), the breed is well known and the meat is considered far tastier than that obtained from many other breeds. As a consequence, a Mangalitza joint from a Hungarian butcher will cost quite a bit more than one from the average farmyard pig. Stock has also been exported to Spain and Portugal where the breed is popular due to the fact that it produces something similar to Parma ham once it has been air-dried and cured.

The extraordinary woollen coat is believed to protect them from the extremes of the cold and wet in winter and the heat in summer. There are three accepted colours: the red (rather than being a real red, most are similar in colour to the Tamworth); the Swallow Belly (black is the prime coat colour, but the belly is white like the breast of a swallow. Interestingly, In parts of Hungary and into Slovenia, the black back actually has a reddish tinge) and the blonde (which varies from a creamy hue to true blonde). As well as the accepted colours, pied and other unusual markings can sometimes be seen.

It is very slow to mature (which is of no consequence to the hobbyist or seeker of a very unusual pet pig breed), easily handled and, unsurprisingly given their origins, very hardy. Although litters are likely to be smaller in number than mainstream breeds, the sows make very attentive and careful mothers.

LEFT Mangalitzas lose some of their coat in winter.

OPPOSITE A young Mangalitza with an excellent coat.

MICRO-PIGS

Anything out of the ordinary seems to attract curiosity; no wonder then that recently much attention has been paid to the arrival of the Micro-Pig. Vietnamese Pot-Bellied and Kune Kune pigs (see separate entries) have both suffered in the past from being a fashionable pet with a purchase price to match demand. Fortunately for both breeds, today they are mainly in the hands of knowledgeable owners. Now it seems that it's the turn of the Micro-Pig.

Micro-Pigs are the miniatures of the porcine population and even when fully grown, might stand only about 50cm (20in) high. They have attracted the attention of celebrities from film actors to footballers, with the result that they have been made much of in the media and there is apparently such a demand that breeders have waiting lists, with individual animals fetching record prices. Once all the excitement has died down, they might, however, make a suitable alternative for someone who wants pigs as companion pets, but who does not have much space to hand.

A Micro-Pig is certainly not a pure-breed and those most frequently up for sale occur as a result of several generations of careful (some might say, irresponsible) breeding from the smallest of domestic pigs such as Tamworth and Gloucester Old Spots, together with miniature Vietnamese Pot-Bellied pigs. As a result, it is impossible for an actual size and weight to be guaranteed because much depends on the parents. There is however, a genuine Micro-Pig that will breed true. The Göttingen mini-pig was developed in Germany in the 1960s for the sole purpose of supplying the medical industry with an animal that could be used in laboratory conditions to monitor drug developments related to pharmacology and toxicology. Two of the biggest producers of the true breed can now be found in America and Denmark, with genetic consistency being regularly monitored by the University of Göttingen.

Like the Micro-Pigs rapidly becoming so popular as pets, Göttingen mini-pigs reach adult weight at around two years of age, but sexual maturity occurs much earlier. As with all breeds, mini-pigs should be kept in groups and, despite the media hype about them being possible to house train, they need to have access to the outdoors in order that their natural instincts can be indulged.

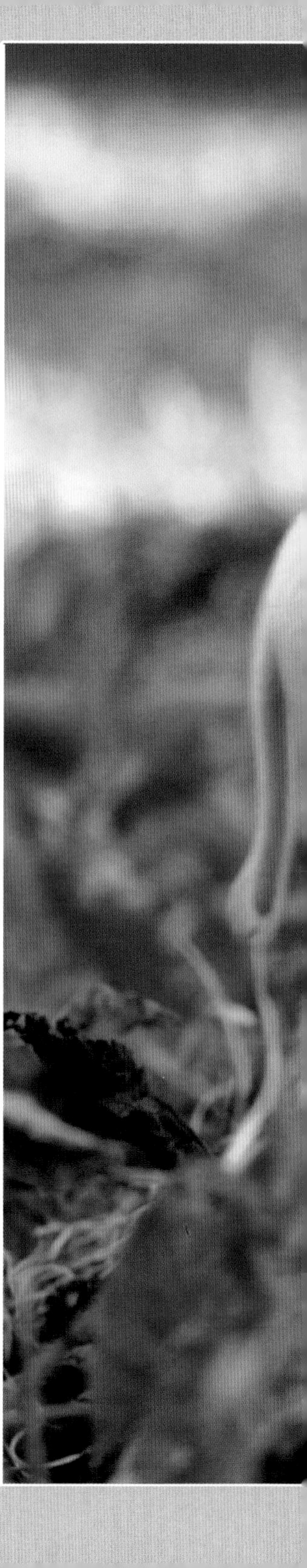

RIGHT Cute and appealing, Micro-Pigs are currently enjoying a great deal of popularity.

MIDDLE WHITE

Apocryphal though the story may be, common thinking has it that this particular breed originated as a result of some confusion over sizing standards at an agricultural show in Yorkshire during the 1850s. It appears that the stewards of the day could not decide as to whether the white pigs belonging to one Joseph Tuley should be included in the Large White or Small White section and so, rather than eliminate what looked to be extremely promising animals, a third classification was created.

Whatever the truth of the story, the Middle White definitely originated in Yorkshire and almost certainly came about as a result of selected and careful mating between individuals from already established herds of Large and Small Whites. Because of their origins, the breed type is known in parts of America as Middle Yorkshire.

Because of its size and ability to do well in almost any situation, it very quickly turned out to be popular with backyard pig-keepers everywhere, especially in cities (it was at one time, given the nickname of London Porker). Unfortunately, in Britain at least, the numbers began to decline not long after World War II, perhaps because of the fact that features once considered desirable (small, slow-growing producers of exceptional belly pork and fatty bacon) were no longer wanted by an increasingly health-conscious public.

Although it is slowing regaining its popularity, especially among small-scale pig-keepers and hobbyists because of its placid nature, easy handling and superb parenting qualities, it is still one of the rarer breeds and should be considered carefully as a possible breed by anyone going into pig keeping for the first time.

Like almost all traditional breeds, Middle Whites are slower to mature than commercial breeds – a point worth considering if meat is your main objective – however, most lovers of the breed think that the wait is worth it. Interestingly, there are statues to this breed in Japan, as they think it is the best thing going!

The Middle White has an upturned snout, short dished head and disproportionately large ears which, although pricked, to tend to fall slightly forwards – the overlarge ears and dished head when seen from the front have caused some people to compare it to the head of a vampire bat. The upturned nose might make the breed worth considering in places where you'd rather not have your paddocks turned into ploughed fields, as it is believed that most breeds with such a nose are grazers rather than rooters. Having said that, it must be pointed out that no pig grazes neatly, efficiently and in the same way as a cow or sheep, so don't expect there to be absolutely no damage to your grassland.

RIGHT A Middle White boar exhibiting the typical facial expression that causes many people to compare them to vampire bats!

MULEFOOT

It being such a rare breed, some might think that this very unusual and extremely interesting type of pig should perhaps not be included in a section, the main purpose of which is to help the would-be pig-keeper in selecting and sourcing a breed suitable to their particular circumstances. It is however, imperative to describe at least one such breed in order to emphasize the fact that some breeds might die out without the intervention of enthusiastic devotees. Remember that it's not only the likes of wild tigers, for example, that are in danger of extinction – some domestic breeds of livestock are too.

Rare and interesting because of the fact that their hooves are not cloven like most pigs, but more resemble the feet of a mule, this particular breed has its origins in America where it has a long history of agricultural use. Mulefoot numbers are classified as 'critical' by the ALBC, but the situation has slowly improved over recent years to the point where it is believed that there might now be as many as 600 individual examples of the breed.

Mulefoots should be all-black in colour but occasionally one may be found with white spots. Although undoubtedly American, the area in which it was actually developed is not known for definite, although FD Coburn, writing in his book *Swine in America*, published in 1916, states that the Mulefoot hog was found in Arkansas, Missouri, Iowa, Indiana, across the south west, and in some parts of Mexico. Although it is a big animal, it is very gentle and if space permits (and a supplier could be found) it might be worth considering by any American readers.

BELOW The Mulefoot has a hoof more like a donkey than a pig – hence its name.

OXFORD SANDY AND BLACK

Thoroughly recommended and often described as being the perfect first-timer's pig, the Oxford Sandy and Black is arguably one of the most attractive of all pig breeds. In addition, they have a reputation for doing well under most systems, but especially those of the smallholder or hobbyist. They are good foragers and therefore economical to keep and, due to their coat and colour markings, are less likely to suffer from the sun than some other breeds.

Breed standards indicate that the main body colour can be anywhere between light sandy through to a deep rust. Correctly, the random black markings should be more blotchy than spotty and the feet white, with a white blaze on top of the head. The head should be strong in appearance, straight or slightly dished, with lop or semi-lopped ears. It is one of only a few breeds (the Kune Kune being another – see separate entry) to have tassels attached to the lower jaw.

The breed has had a chequered history and at one time numbers dropped so low that there was a very real possibility that the OSB (as it is sometimes known) could have become extinct. In fact, as is often the case with minority breeds in both the poultry and pig worlds, it may well have done were it not for the dedicated efforts of a handful of knowledgeable enthusiasts, which would have been a great shame for a pig breed that is thought to have been around for 200–300 years.

ABOVE Although the OSB is undoubtedly old in origin, it wasn't until relatively recently that this perfect smallholder's pig had its own breed society and a herd book.

PIETRAIN

Originating from Belgium, the Pietrain is reasonably well known in several countries – including Germany, Spain, France, Britain and America – as a result of it being exported during the early 1950s for cross-mating with other breeds in the hope of creating a lean meat-producing hybrid, at a time when most other breeds produced a fatty meat. The Pietrain is reckoned to have as much as 66 per cent less fat than some breeds and the meat is said by many to have a sweeter taste.

There is occasionally a problem when the breed is kept in commercial units and the BPA explains that, despite it being an excellent pig when thus used, very high yields of lean meat are sometimes, 'associated with the presence of the halothane gene for Porcine Stress Syndrome', and 'for this reason, the use of pure-bred Pietrain in British pig production is relatively rare.' However, a number of pure-bred herds are maintained to supply stock for cross-breeding programmes and, importantly, by a dedicated band of enthusiasts who breed and exhibit Pietrains purely for the pleasure that they give.

It may need more care in the colder months than some other types of pig but on the plus side, most strains are quiet and easily handled. It is a medium-sized, prick-eared breed and is white with black spots – although around the spots can be found small rings of light skin pigmentation that results in white hairs. Typically then, Pietrains are sometimes described as being piebald in colour.

BELOW Around the spots of a Pietrain can be seen a paler ring of colour.

SADDLEBACK

There used to be a Wessex Saddleback, originating in Dorset, which was a black pig with black legs and a broad white band circling the main part of its body. There was also an Essex Saddleback, similar in all respects except that the white band was set further back on the body, resulting in it having white back legs. In 1967, the decision was taken to amalgamate the herd books of both in order to form the British Saddleback, which is considered a shame by some as they had such obviously differing origins. However, the merge was done for all the right reasons, namely to introduce hybrid vigour and to minimize the possibility of undesirable hereditary faults as a result of too much inbreeding among the few herds still in existence.

The white band that circles the body of a Saddleback varies in its width – even within members of the same litter. Individuals might have markings that range from a width almost as wide as the body length of the pig, right down to the other extreme where very little white is evidenced. However, British breed standards state that the belt should include the forelegs and, while white markings are permissible on the back legs, they should reach no higher than the hock. White is also allowed on the nose and the tip of the tail. The Saddleback is a perfect pig for the beginner as it is quite hardy, smallish, docile and a good mother. It is also a good grazing animal unlike some other breeds, which seem intent on rooting to the centre of the earth even when given large expanses of ground on which to roam.

ABOVE The Saddleback has many good qualities, particularly its excellent temperament and mothering abilities.

TAMWORTH

Tamworth is a town in the English county of Staffordshire and, unsurprisingly, this particular breed originates from that area in the days long before it became a conurbation, when rural dwellers were given permission by the landowner to allow their animals to seek out their own food in surrounding woodland (a practice known as 'pannage').

Easily recognized by its colour, which ranges from a lovely golden-red to deep russet, the Tamworth still has the look of an ancient pig and it's easy to imagine them rooting about the heaths and forests of England in the days of Robin Hood! Long and lean, it has quite a streamlined appearance when compared to other breeds and has noticeably longer legs, neck, head and snout – all factors that correctly suggest the breed was developed as an animal designed to live without artificial shelter and to wander long distances during its foraging forays.

Their inherited tendency towards roaming for their food might make them less suitable than some other breeds for consideration by the hobbyist but, if you can provide them with a small wooded area in which to forage, they will be kept amused. They are often used to clear such places of unwanted scrub and bracken, especially on sloping ground where no other forms of clearance may be practical. Pigs in general, but Tamworths in particular, are brilliant at eradicating bracken as they bruise the plant (by treading and breaking mature stems), cut the young fronds by grazing and plough up the area by rooting, often leaving roots and stems on the surface, which will eventually die after being exposed to the sun.

The ears are erect and pointed, and its coat is often longer and coarser than many other breeds (most likely evolving in order to protect its owner from the natural elements). As with most animals, the Tamworth sheds its coat periodically and so in the summer when the hair is less pronounced, it is quite partial to wallowing in mud, which protects it from the harmful rays of the sun.

The Tamworth is generally a hardy pig but it still requires protection from the elements and should be given sheltered accommodation in the same way as any other breed. Due to their ancestry including wild boar, Tamworth males have been known to develop long and potentially lethal tusks as they get older and these could well require periodic trimming – for that reason alone, the inexperienced pig-keeper may be well advised to start with gilts.

RIGHT If you want a breed that will clear overgrown woodland, the Tamworth is an obvious choice.

VIETNAMESE POT-BELLIED

A decade or more ago, these pigs had a cult following – they appeared as pets in the most obscure of places and were owned by the most surprising of people. As one celebrity owner recounted in an article written in 1997, 'Angel travelled to us by private jet, transferred to the back of a Porsche and was chauffer-driven to our door. We heard her long before we saw her and when her crate was opened, out rushed an adorable black bundle of joy ... She became a friend immediately, copying the dogs, going for a walk, coming when called (if convenient), and sitting on the sofa or bed. We decided her quality of life would be enhanced if she could get into the car... Angel, now mobile, undertook the school run, visited the supermarket (car park only), the vet and my mother for lunch.'

More usually however, Vietnamese Pot-Bellied are content with a less glamorous lifestyle and thrive far better outside where they can behave as pigs are meant to. They need to graze, root about in vegetation and wallow every bit as much as any other breed and, like all pigs, prefer the companionship of at least one other porcine friend.

Vietnamese Pot-Bellied are smaller than most farmyard breeds and are easily identified by their sunken-bellied shape, dipped back and wrinkled face. When first introduced from the Far East, their usual colour was black, perhaps with a patch or two of white and occasionally with a pink belly. In the intervening years, breeders have developed many other colour variations including red, silver, tan and white. The hair is very coarse and sharp and the boars grow tusks that may require trimming from time to time (they also produce sperm earlier than some more conventional breeds so care needs to be taken when keeping young stock of both sexes together).

When kept as solitary pets, it was noted that, on maturing, the Vietnamese Pot-Bellied could become quite territorial, especially if a stranger came near its bed area. Conversely, when kept as a pig should be, they are known for their generally gentle nature and there are those who claim that it is one of the most intelligent breeds of pig in existence. It is a sobering thought that, according to Lyall Watson writing in his book *The Whole Hog*, the entire New World stock originated from just 18 pigs imported from Sweden, but mainly as a result of the pet fad, 'there are now something like 50,000 ... trotting about ...' Thankfully, the fashion of keeping them as domestic pets has now run its course and so, should you fancy the idea of Vietnamese Pot-Bellied, there is an extremely good chance that any you see for sale will have been bred by knowledgeable and experienced enthusiasts who will be able to help and advise you.

LEFT A relatively small pig, the Vietnamese Pot-Bellied reaches puberty earlier than many breeds.

OPPOSITE Vietnamese Pot-Bellied are easily identified by their sunken bellies and dipped back.

WELSH

The Welsh is smaller than many breeds and is, for that reason alone, well worth considering as a hobby pig where space is limited. Because of its size, it is perhaps not best suited to being included with a mixed herd, as there is always the danger that it may be continuously pushed away from the food by its larger neighbours (though an extra trough or two might alleviate the problem).

As recently as the beginning of the last decade, the pedigree Welsh had almost died out in the UK, due in part to the fact that its meat was considered too marbled (fatty) for modern health-conscious tastes. Fortunately, more recent trends whereby consumers are encouraged to 'think local' and go back to traditional meats and methods, have helped the Welsh breed a great deal. In its native country, the breed has been further assisted by funding from the government in order that the Pedigree Welsh Pig Society (PWPS) might be formed – resulting in a directory of over 70 breeders being created in a 12-month period (2009–10).

The Welsh has been known since at least the 1870s when it was often bred and taken into Cheshire to fatten on the whey and other by-products created by the dairy industry. The breed's popularity increased even further around the period of both World Wars, the first Welsh Pig Society being formed in 1922, about the same time as the first pigs were being exported to America and elsewhere. Eventually, the breed was represented by the National Pig Breeders Association (now the BPA), under whose 'umbrella' it continued to flourish – partly because of the recommendation of the 1955 Howitt Report (see Landrace entry), which outlined ways in which Britain could compete with the Danish, Dutch and Irish pig industry.

According to the PWPS, one George Eglington paid the then enormous sum of 4,500 guineas for three Landrace gilts and a boar, which he then used to improve his existing herd of Welsh, the offspring of which were then bred with more pure Welsh and subsequently became the foundation stock for the modern Welsh breed. Commercial interest in the breed declined during the 1980s but, hopefully, recent enthusiasm for this white, lop-eared pig (described as being pear-shaped when viewed from the side or above), should help to redress the balance. It seems that it has, for although in 2005 it was declared an 'endangered' species by the RBST, it has recently been reclassified as 'rare'.

LEFT The Welsh breed is enjoying something of a revival in popularity but despite this it is of 'rare' breed status.

OPPOSITE Typically pear-shaped, the Welsh is a wonderfully docile breed and the sows make excellent mothers.

RARE BREEDS

Part of the attraction of choosing a suitable pure-breed is the fact that you are helping to conserve all-important bloodlines and long-established strains. Should circumstances and personal preference allow you to keep one of the recognized 'rare' breeds, then so much the better as far as future generations are concerned.

In the UK, the Rare Breeds Survival Trust (RBST) based near Kenilworth in Warwickshire, currently classify eight breeds as being of rare breed status: namely, the Berkshire, British Lop, British Saddleback, Gloucester Old Spots, Large Black, Middle White, Tamworth and Welsh.

Understanding exactly what is meant by 'rare breed' is, however, not that straightforward. In the UK, for example, it might be that, rather than being rare in overall numbers, a breed is rare in its geographical concentration – as the RBST points out, 'some breeds may be numerous but if the majority are found in a small geographical area the breed will be highly vulnerable to disease epidemics.' Their categories of 'critical', 'endangered', 'vulnerable', and 'at risk' are therefore, more likely to refer to this factor rather than that individual pig numbers are in danger.

In the US, things are somewhat different and the American Livestock Breeds Conservancy (ALBC) Conservation Priority List classifies within its parameters, rare pig breeds as being numerically challenged. 'Critical' would, in this instance, suggest that there are fewer than 500 breeding animals, five or fewer primary breeding herds and also that the breed is endangered worldwide. The 'watch' category is used to indicate breeds containing fewer than 5,000 breeding examples and also those with genetic concerns or – as in the UK – that there is a limited geographic distribution. Pigs classified as 'recovering' include breeds, 'which were once listed in another category and have exceeded 'watch' category numbers but are still in need of monitoring.' On its list, the ALBC include the Choctaw, Gloucester Old Spots, Guinea Hog, Large Black, Mulefoot, Ossabaw Island, Red Wattle, Hereford and Saddleback.

It is as well that these organizations exist, not only for the future wellbeing of pig breeds, but also for the less popular types of poultry, cattle, sheep and horses. Anyone with any interest in domestic stock should therefore seriously consider joining one of these groups in order to offer financial and moral support.

BELOW The Ossabaw Island breed has strong genetics but it is rare and is considered 'vulnerable' by the ALBC.

SHOWING PURE-BREEDS

In addition to visiting shows in order to gain a better idea of which breed of animal will suit your requirements the best, a pig show (usually held either as part of a large traditional agricultural show during the summer months, or possibly as a winter event organized by a particular breed club) is probably the finest chance you will ever have to really appreciate the individualism and character of a pig.

No matter how well trained they may be at walking around the ring there is almost always something uncertain about the eventual outcome! Whereas with goats, sheep and horses, an animal can be led by a halter and trained to stand almost on command, there is rarely such luxury afforded to the porcine exhibitor – a pig will go in the general direction required but you can almost see it thinking, 'well, if I must, I must', a fact it often voices with a series of grunts. And that is not to say that it objects to the experience (pigs are notorious for loving attention and the limelight), merely that it is making its opinion felt!

There is a skill involved with showing pigs that is beyond the remit of this book and the experiences of a well-respected breeder and handler will help far more than any amount of written words. Basically, however, a show animal obviously needs to conform to the breed standard and to show a general degree of soundness. It must be washed, groomed and possibly even had a hair-trim around the tail and the inner ears. There should be no dirt onthe feet and certainly no sign of litter debris anywhere on its body as a result of it being penned while waiting to go into the show ring (for that reason, it is a sensible precaution always to include a stiff-bristled brush in the pocket of your white show coat).

One aspect of showing that remains the same – and equally applicable to whatever animal you are exhibiting – is always to keep your pig between you and the judges: if you stand on the wrong side it prevents them from seeing the animal's best profile, and after all, it is your pride and joy that is on show, not you!

ABOVE With the aid of a cane or slapstick and moving board, it should be possible to encourage your pig around the show ring.

CROSS-BREEDS AND HYBRIDS

If you are intending to exhibit your stock, then it will be necessary for your pigs to be pure-breeds. Cross-bred pigs might, however, hold an attraction for some, especially if it comes to providing a meat carcass.

One breeder was recently quoted as saying, 'The traditional breeds tend to carry a bit more fat round the meat. That's how people used to like it – they didn't get fat on it themselves because they were mostly out working a 12-hour day ...' Nowadays, if you are looking specifically for a meat animal, it is possibly better to go for a cross and not let them get too big before taking them to slaughter.

The difference between a cross-breed and a hybrid needs to be explained. A cross-breed is, to use an expression that most will understand, a 'Heinz 57' – a mish-mash of animals bred together throughout generations. It might, therefore, be a beast that has a particular pairing in its ancestry, which has then gone on to be bred with totally different breeds until such time that no one has any idea as to its lineage.

A hybrid, on the other hand, is more likely to be a combination of two parents brought together for a particular purpose (perhaps to eradicate certain undesirable tendencies found in the pure-breed), the offspring of which is known, from experience, to do better in whatever sphere. To produce the same again, one would have to go back to the parental line. If you crossed a hybrid with another hybrid, you would eventually end up with a cross-breed.

Both hybrids and cross-breeds might be an option, but not before giving careful consideration to the traditional and/or pure-breeds available. Remember that perhaps the greatest advantages pure-breeds have over cross-breeds or even hybrids is the fact that it is possible to predict their adult size, conformation and likely behavioural characteristics, all of which are important factors in deciding whether or not they will be, to use modern parlance, 'fit for purpose'.

BELOW Commercial hybrid pigs on free range.

EXTINCT BREEDS

Although many traditional breeds – sometimes classified as 'rare' in the US, Europe and the UK – have just about survived both the commercial breeder's whim and government edict, others have, quite literally, gone the way of the dodo. Some bloodlines are possibly included in today's breeds and, diluted though they may be, the genetics of semi-wild breeds of the Middle Ages still remain in breeds such as the Hampshire, Berkshire and Gloucestershire Old Spots.

ABOVE Although the Mangalitza is Austria-Hungarian in origin, it has a direct genetic link to the now-extinct Lincolnshire Curly-Coat breed.

As well as the Lincolnshire Curly-Coat mentioned in the Mangalitza entry, other UK native breeds driven to extinction include the Cumberland, Ulster White, Dorset Gold Tip and Yorkshire Blue. Sounding more like names given to traditional cheeses, they are nevertheless a sad indication of some insular thinking on the part of commercial pig producers of the past who were, in the main, merely concerned with producing an animal that could withstand intensive housing and a carcass that could be sold to the butcher (and subsequently the consumer) in the least amount of time.

Although Britain is a country on its own, even within its shore restrictions there have been further boundaries. With no means of movement other than on foot, a pig most suited for the area most likely stayed in the area. Despite this, the various English pig breeds have, irrespective of some tragic casualties, more or less survived, even if only as a result of them being exported abroad. The Welsh breed is undergoing a recent revival (see breed entry) and Ireland has always been a place where pigs of one breed or another have always been kept in a rural community. The Highlands of Scotland have fared less well: crofter's pig breeds such as the Grice – a small hardy breed originating from the Shetland Isles – was completely bereft of bloodlines affecting modern-day breeds and became totally extinct in the 1930s.

In America, care must be taken that the Mulefoot (see breed entry) and a few other minority breeds do not go the same way.

BUYING AND SOURCING

BUYING THE RIGHT TYPE OF PIG FROM A REPUTABLE SOURCE IS THE MOST IMPORTANT THING YOU CAN DO TO MAKE YOUR VENTURE SUCCESSFUL. HAVING CAREFULLY CONSIDERED THE RIGHT BREED, EQUALLY AS IMPORTANT IS THE NEED TO BUY HEALTHY, WELL-BRED ANIMALS. IF YOU DON'T, IT IS LIKELY TO LEAD TO DISAPPOINTMENT, A WASTE OF FINANCES AND A FAILED ENDEAVOUR. BUT IF YOU DO, YOU WILL ASSUREDLY BE ADDICTED TO PIG KEEPING FOR LIFE.

Solutions Ltd
07976 255 220
Ltd

HOW TO BUY THE BEST

It is important to find a respected seller, as a well-known breeder has a reputation to maintain. When you have, go and see them (taking with you a knowledgeable friend if possible) and try to ascertain that their stock is suitable for your particular purpose and, if so, that the examples of what are on offer have been kindly looked after – animals thus treated generally have a far gentler temperament. If it's young stock you are after, ask whether it might be possible to see any adult direct relatives of those you are considering buying, as this will give you a good indication as to how yours will eventually look.

But how do you find such a paragon? There are many breed associations with excellent websites on which you will find the contact details of well-respected breeders. The associations might be national or extremely local, but all will offer the opportunity to be put in touch with experienced and enthusiastic suppliers who will almost certainly do their utmost to help you at the initial stages, as well as being prepared to be at the end of a telephone line if you require advice at a later date.

SHOWS, SALES AND MARKETS

Because of the need for movement licences, it is unlikely that you will visit a show or sale and buy pigs on a whim (unlike rabbit and poultry shows where casual visitors can look, succumb and buy – irrespective of whether or not they have anywhere suitable to keep the unfortunate creatures). Instead, it is likely to be a conscious effort – you will know what you are looking for and will have already identified the necessary legalities involved with such a purchase.

Rare breed sales are generally an excellent and safe way of buying stock, as they are normally organized under the auspices of an accredited rare breeds society that will only permit bona fide sellers to participate. Alternatively, should you be merely searching for a pen of healthy commercially produced weaners – the actual breeding 'pedigree' of which does not matter to you – then your local weekly or monthly livestock market might be the place to go.

ONLINE SOURCES

A somewhat less reliable method of sourcing your pigs is via the various pig-related blogs on the Internet – and here it must be made clear that this is not intended to include established breeders with their own websites, but more the average 'punter' with the odd litter of pigs to dispose of. Although you may be lucky, there is always the danger of buying inferior stock, which may also be misrepresented in the online advert. If, for example, a seller claims that his or her pigs are from a well-known line and that certain herd-book sires feature in the breeding, but that unfortunately, 'no papers are available', how can you be sure that you are not, quite literally, buying 'a pig in a poke'?

THE FASHION TRAP

Periodically, and for no apparent reason, the most unlikely people seem to fall in love with a particular breed of pig, which then develops a 'cult' following. In the past, these have included the Kune Kune and the Vietnamese Pot-Bellied. The current 'pigs of the moment' are generally known as Micro-Pigs (see chapter 2) and, while there is nothing wrong with keeping such an animal – in the correct way – there is always the danger that unscrupulous breeders will attempt to jump on the bandwagon. *Caveat emptor* – 'let the buyer beware'.

ABOVE RIGHT An auctioneer talks up the finer points of a breeding boar while potential buyers look on with a practised eye.

OPPOSITE Because of their ability to do well in almost any situation, pigs were even popular with city dwellers in bygone eras, as can be evidenced from this market in an industrial setting.

A PIG IN A POKE

Despite being fully understood in modern everyday language, the origin of the expression 'a pig in a poke' is, nevertheless, quite old – perhaps over 500 years. A 'poke' was, at that time, a hessian or canvas bag, derived from the French word *poque*, which was used by artisans to describe a bag in which they kept their tools. The size of the poke meant that it was ideal for carrying a young piglet home from the market. Unscrupulous traders would turn this means of transport to their advantage and let the buyer choose their pig from a litter they had for sale. Turning their back on the purchaser so that he couldn't then see what was happening, they would substitute the chosen animal for the runt of the litter and quickly place it in the poke – thus leaving themselves with the best piglet to sell again. The buyer would have been unaware of the trickery until reaching home.

WHEN TO BUY

The infamous 'pig cycle' is an artificial cycle of prices induced by market forces and popular demand. In essence, pigs become profitable (or certain breeds more fashionable) so more producers keep them, the market weakens and producers stop keeping them, eventually the price rises again and the cycle continues. By observing these fluctuations you may be able to purchase stock at a better price.

For meat production, there are slight seasonal variations. Pork for example, has marginally better returns during winter months and hams are also at a premium around Christmas. Therefore, the price of your young stock for rearing on could well be higher some months before these periods. Even when your main consideration is simply to supply your own freezer, it will still pay to watch the market and purchase whenever it seems favourable.

Traditionally, pigs were always bought with the intention of killing in late autumn when the cooler weather allowed the meat to be cured, by salting or smoking and when temperatures were lower and flies less populous. With modern-day living and easy access to refrigeration and freezing, it makes little difference.

Perhaps your main consideration is your annual holiday: a good neighbour or family member might be persuaded to look after your long-term stock, but while asking a neighbour to feed a few hens or a cat is not too difficult for them, asking them to be responsible for your precious pigs may be asking too much. Buy your initial stock after the holiday and not before – this will give the longest opportunity for you to get to know your pigs and for them to get to know you. If your main interest is likely to be in rearing pigs for the table, work it so that your animals are able to be despatched and in the freezer beforehand, leaving your housing cleaned and empty for your return and a new intake.

A final yet very important consideration is that your feed costs of rearing will be a little higher in the winter as the pig will use more food just to keep warm and you may not have the benefit of surplus garden produce to supplement their rations – in which case, the ideal solution might well be to purchase your stock in the spring.

BELOW Spring-purchased piglets have the advantage of warmer weather and the opportunity for some natural foraging.

AGE AT WHICH TO BUY STOCK

Buying stock at a young age has many advantages, much of which, it must be admitted, appertains to cost; but add to that the fact that you will be able to introduce them to your own particular system with the least amount of stress. Piglets are very adaptable and will soon learn, while an older animal may have already adopted some habits that might be difficult to break.

For meat-rearing projects (in a situation where you do not want to buy the whole litter), it might be possible to obtain the smallest pigs from a litter at a good price. While you may not be as fortunate as he was, one of the authors once managed to obtain two such pigs for next-to-nothing as they were being bullied by their litter mates.

As regards your potential breeding stock, it is a far better option to buy at a more mature stage: a delightful looking piglet may turn out to be rather less appealing when mature, and if you are looking for a true representation of a pure-breed for pleasure and for show (where obviously a well-shaped and gentle animal is preferred), such a purchasing route is almost essential. In essence, you are buying time because instead of waiting for six months for your animal to mature, you may be able to begin with mating straight away.

All in all, it is very much a case of, 'you pays your money and you takes your choice', and much depends on whether you are keeping pigs for meat or for hobby purposes. Alternatively, you might decide to combine the two previous suggestions and buy a sow with a litter – which will eventually provide you with a breeding animal as well as youngsters, some of which you can sell on while keeping a couple back for the freezer. Another option as regards age is to consider the purchase of adults that have already produced or sired a litter – in which case they will be at least a year old. They will be more expensive than any of the other options but at least you have the security of knowing that the animals are fertile and 'proven'.

ABOVE The age at which you buy your pigs will be reflected in the price you pay.

POINTERS TO PURCHASE

Whatever stock you intend to buy, the rules of common sense apply. For example, a quick visual inspection of the premises will tell you much about how the animals are being raised. Beware of a ramshackle collection of buildings with leaking roofs and draughty walls. Untidy piles of manure scattered about the yard betray a poor system of husbandry.

Apparent health is, of course, an important issue; a pig should look alert and interested in its surroundings – in fact it would not be too fanciful to suggest that it ought to have a twinkle and element of mischievousness in its eyes! Both front and back legs should show no sign of lameness (remembering that most breeds of pig walk with a certain 'stiffness') and there should certainly be no signs of an obvious ribcage or, conversely, too much fat (most noticeably around the neck).

Many pig-handlers will advise you to check the overall size and shape of the belly, which should be rounded when viewed from the side rather than drawn upwards into the body. They might also suggest that you look from the back of the pig in order to check that both sides look the same and have no unusual swellings.

PIGLETS

If it's piglets you're after, you should be shown a litter from which to make your choice. Beware the shy, quiet animal sulking in the shadows – it may be fine and just having a bad day, but the risk is not worth it. Look for the ones that come forwards with confidence; pigs are inquisitive animals and if they have been kindly dealt with, they will always come to see you. Look for a healthy coat, they may have a little dirt on them, but nevertheless, it should show the normal cleanness and freedom from parasites. Piglets, like any animals, should have clear, bright eyes without any matter in the corners and the ears should appear clean inside. Generally, any tag or clip-mark should be free from redness or swelling.

BREEDING SOWS

If it is a mature female you intend to purchase for breeding, examine her tummy – she may complain at being unceremoniously handled but will get over it. Look for two tidy rows of nipples and make sure there are sufficient – they can vary but 12 to 14 is normal. On a more mature female, handling may not be so easy, but even so, a general inspection can be made. While she is standing, run your hand under and along her tummy to check her 'milk bar'; if she is the sort of animal you want, she should not show discomfort and may even enjoy the experience! While she is being handled, check the rear end under the tail for any apparent abnormalities and never purchase a female for breeding that has had her tail docked – the tail serves a purpose to keep away irritating and biting flies.

FURTHER OBSERVATIONS

Temperament is all-important – a nervous or seemingly aggressive pig is not one you want to take home with you, especially if you are intending it to become a long-term happy member of your family. Apart from never being able to trust such a beast, it will undoubtedly prove difficult to catch, examine or treat whenever such things might be necessary.

Watch your potential pig for a time and observe what it is doing. Over a period of 20 minutes or so, you should see it eating, rooting and drinking. Don't be embarrassed to check that it urinates clear rather than cloudy liquid or that its faeces are solid rather than diarrhoea-like.

If it happens to be a sow with piglets and the youngsters feed quietly and without too much fuss and bother, they and their mother are almost certainly healthy (a feeding sow may well make gentle grunting noises as she lets down the milk). As a general overview, look out for any unusual wounds – a pig might have a healed gash as a result of being inadvertently slashed by the hooves of another or even have a small rupture from birth, but otherwise they should be reasonably blemish-free.

OPPOSITE A good clean profile generally indicates a healthy pig.

BELOW Observe your potential pig before purchase – if it's healthy and well, it will probably be observing you!

ARRANGING TRANSPORT HOME

Having purchased your pigs and filled in the necessary movement licences, the next step is to get them home. If the journey is likely to be a long one, you must plan your travel arrangements in such a way that you make provision for the animals to drink, feed and rest without having to endure continuous momentum.

In the past, it was a simple matter of placing youngsters in a loosely woven hessian sack, or adults into the back of an estate car or small van, with a generous layer of straw or wood-shavings on the floor. Ventilation in the form of a semi-opened window and a gentle ride over the worst of the bumps would have been all that was necessary to get your pigs home. However, this method of transport is now considered wrong under current legislation and should only ever be undertaken in an emergency situation.

For a larger animal, you may be able to negotiate delivery as part of the deal if the breeder has suitable transport. However, in the long term, you will need to consider having similar arrangements of your own. By far the most sensible option is a trailer. While a horsebox-type trailer is more than adequate, it will be expensive and may require a towing vehicle larger than the one at your disposal. Many manufacturers produce trailers specifically made to transport pigs and sheep. They are usually constructed of lightweight materials, completely covered over in hard material and fitted with a ramp for loading and unloading. Quite often they also include a side access door and have the advantage of being easier to reverse, something that many people find difficult and require a little practice at.

The same trailer will be very useful for collecting bedding and food from your local merchant; they are easy to clean with a standard hose or pressure washer and with care will last a lifetime. This type of trailer can easily be pulled by the normal family car fitted with a ball-type tow hitch, but do check the towing weight recommended in your vehicle handbook.

BELOW If you have your own trailer, transporting your pigs will be far less stressful if you've previously accustomed them to going up and down the ramp.

SETTLING INTO A NEW HOME

On arrival, your pigs will be keen to get out and stretch their legs, so the new home should be ready and waiting. If it is a sty, the sacking should be down over the house door, or the top half of the stable door closed down to keep it a little dark inside. A good deep bed of straw will encourage the pigs to forage about and eventually lie down to sleep.

It is a good idea to sprinkle a handful of pig nuts in the straw to keep them occupied; failing that, an apple sliced up and so distributed will please them and of course, they must have clean fresh water. As their first 'toy', consider including a short length of tree branch or a log or two for them to play with (see chapter 7). Unless they have not been fed for some hours, it is unwise to feed them too much at this time, as they may be a little stressed from the journey. When their first feed is due, make it a small one and observe if they are eating and defecating normally before giving a normal-sized meal the next time.

If your choice of housing is a traditional sty and pen, then there is no need to keep them enclosed within the house section. If, however, it is an ark within a large fenced area, it might be as well to organize things so that your pigs arrive in the early evening and can be locked in overnight before being given the chance to explore their surroundings first thing in the morning. Whatever, last thing in the evening, quietly stand outside their quarters – if they are sleeping soundly all is well.

USEFUL ADDRESSES

In addition to the association addresses included here, remember that many of the traditional and/or rare breeds have their own particular society, the contacts for which will be found using an Internet search engine.

USA

United States Department of Agriculture (USDA)
1400 Independence Avenue SW, Washington, DC 20250
202-720-2791 www.usda.gov

American Farm Bureau Federation (AFBF)
600 Maryland Avenue SW, Suite 1000W , Washington, DC 20024
202-406-3600 www.fb.org

American Livestock Breeds Conservancy (ALBC)
PO Box 477, Pittsboro, NC 27312
919-542-5704 www.albc-usa.org

National Swine Registry (NSR)
PO Box 2417, West Lafayette, IN 47966-2417
765-463-2959 www.nationalswine.com

UK

Department for Environment Food and Rural Affairs (Defra)
08559 335577 (from outside UK +44 20 7238 6951) www.defra.gov.uk

The British Pig Association (BPA)
Trumpington Mews, 40b High Street, Trumpington, Cambridge CB2 2LS
01223 845100 www.britishpigs.org

National Pig Association (affiliated to the National Farmers' Union)
Agriculture House, Stoneleigh Park, Kenilworth, Warwickshire CV8 2TZ
www.npa-uk.org.uk

Rare Breeds Survival Trust (RBST)
Stoneleigh Park, Kenilworth, Warwickshire CV8 2LG
024 76696551 www.rbst.org.uk

HOUSING

A WELL-HOUSED PIG IS UNDOUBTEDLY A HAPPY PIG. THERE CAN BE NO GREATER INNOCENT PLEASURE THAN OBSERVING A CONTENTED PIG EITHER LAID FLAT OUT IN ITS BED AND SNUFFLING CONTENTEDLY TO ITSELF, OR ROOTLING THROUGH THE UNDERGROWTH IN THE HAPPY EXPECTATION OF WHAT IT MIGHT DISCOVER. DESPITE BEING MANY AND VARIED, PIG HOUSING IS GENERALLY QUITE BASIC IN DESIGN AND, PROVIDED THAT THERE IS AMPLE SECURE SPACE TO EXPLORE DURING THEIR DAILY PERAMBULATIONS, THERE IS NO REASON AT ALL WHY YOUR PIGS SHOULDN'T BE THE HAPPIEST OF THEM ALL.

COMFORT, SECURITY AND SPACE

The primary consideration as regards housing must, of course, be the needs of the pig. To house it in unsuitable accommodation is asking for trouble. So what exactly does a pig require? Simple: a dry warm bed, a place to rest in the sun, some shade when it is hot, a clean toilet and a good place to eat – not a million miles from our own requirements.

Pig housing must also be secure: pigs can be inquisitive animals and any flimsy workmanship will not last long under their inspection. Any doors must be of substantial, heavy timber, soundly braced, bolted (not screwed), and with good hinges and bolts – pigs just love to rattle a gate to let you know it is time for tea! Fit a top and bottom bolt and ensure the bolts can be fitted with strong padlocks, only a very brave or very foolish thief would attempt to lift an adult pig over a gate!

The traditional sty or 'pig-cote' evolved over many years of pig keeping and appears throughout the world in some form or other: solid in manufacture, it will last for many decades. Typical construction today would be in cement block or perhaps brick. The floor plan should include a bedroom with enough room for two fully grown pigs to lie down and turn around plus the width of the doorway – they will almost always lie out of the direct draught from the door.

The yard area must be big as possible within reason – this will be the area where they eat and defecate and the two should be as far away from each other as possible. The far corner of the yard should slope to a drain attached to a soak-away at least 1m (1yd) square and the same deep, filled with rubble. Never permit this effluent to drain directly into a watercourse. Existing buildings can be adapted, but generally, the roof to the bedroom should be high enough for you to enter for cleaning. Old buildings sometimes have a metal corrugated roof, which is suitable but very cold; consider instead cement fibre or Onduline (bitumen-based) roofing. Alternatively, insulate corrugated tin by covering the roof area with a thick layer of loft insulation (or even straw), before firmly nailing an inner ceiling of plywood.

Where an outdoor paddock is being considered, the moveable ark is, without doubt, the perfect method of housing – especially as it can be placed in the ideal position (see opposite).

BELOW Some commercially available garden-type sheds can make appropriate housing, albeit likely to be temporary.

IDEAL POSITIONING

The prevailing winter wind is perhaps of greatest importance and ideally the house should be faced in such a way that the rear wall is against it and that any openings, such as doors, are not. Second is the subject of shade; pigs like to sunbathe but they can get sunburnt and seek shade when they have had enough. Beware a dip in the ground as this may form a frost pocket and the cold air could well collect in their quarters. The ground should, therefore, ideally slope away in order prevent this possibility.

Access must be considered next; primarily your own access on a daily basis for feeding and cleaning (a half-mile hike can become a chore and you may be dissuaded from sufficient visits in bad weather), but you should also bear in mind that it can be quite useful to have vehicular access to within easy reach of the sty. This may be for a variety of reasons, but mainly because there is a likelihood that your pigs will need to be transported, and the shortest distance from sty to trailer is by far the best. Access to water is required, not just for drinking but also for hosing down after cleaning out. A garden hose can of course be laid from the nearest tap but this may freeze in winter forcing you to carry buckets of water from the house, incurring mayhem and aggravation.

It is well worth considering the positioning of a muckheap. Soiled pig's litter is a valuable asset for the garden and, placed in a tidy heap, mixed with compostable garden waste, will reward you tenfold when it is eventually spread on the vegetable plot. However, an untidy heap or 'midden' is an eyesore to be avoided at all costs. A couple of substantial compost-type bins will keep all tidy and in warmer weather can be covered to keep the flies away. Consider also your neighbours; in fact, consider at all times your neighbours. Not everyone appreciates the sweet aroma and friendly chatter – of the pigs, that is!

ABOVE The ideal location should include a wind break, sunshine and shade.

HOUSING OPTIONS

As far as the pig is concerned, as long as its immediate needs are met, it will not care in what type of house it lives. A roomy, clean, comfortable traditional sty with all the trimmings will be more than adequate; a purpose-built ark is arguably the best bet, but straw shelters have always been popular and if built within a framework, can last a surprisingly long time. It is, however, not really a contest, for there can be no winners, and all options have points in their favour.

However, there are certain advantages to the moveable ark if you have the space. When pigs are kept continuously on the same ground, the land can become quite literally 'pig-sick': worms excreted from their gut breed in the soil and may be ingested later, creating a cycle of infestation that only medication can help to control (see chapter 7). As a means of avoiding such situations, the pig ark is a wonderful invention; heavy enough to provide good shelter, but light enough to be moved from paddock to paddock, they are considered essential by many and cost considerably less than a permanent building.

PIG ARKS

Most arks are built on skids with a tractor hitch at one end to permit them to be dragged from place to place; indeed one of the authors can attest to the fact that provided the ground is firm and reasonably level, two 'good strong men and true' can move them quite a distance – having said that, he did then admit that a tractor is definitely easier! As well as ready-assembled arks – suppliers have been known to deliver on their own transport equipped with an on-board crane, which enables them to swing the ark over fences,

ditches and hedges – it is also possible to buy self-assembly versions.

Modern-day arks are likely to be built with a corrugated iron roof, wooden or polymer ends and some even include a tanalized wooden floor to keep the occupants dry and reduce straw usage. A traditional pig ark is, according to most manufacturers, suitable for eight to ten weaners up to pork weight, a sow and her litter, or two to three adult pigs of a medium-sized breed.

TRADITIONAL PIGSTIES

The traditional stone sty with a yard in front evolved through centuries of trial and error and, according to many pig owners, has never been bettered. If there happens to be one on your property, you are ready to start but you should also remember that they were originally built with a single pig in mind – a situation that should not be repeated today because pigs are gregarious and need the company of at least one other. Irrespective of the need for companionship, pigs nevertheless seem to like this particular method of housing. Provided that there is plenty of bedding in the sleeping quarters, the system is a good way of allowing the inhabitants access to fresh air without letting them root up pasture land. A sty may also prove to be perfect as a farrowing house and subsequent nursery for a pregnant sow. Other buildings can be equally suitable but the insulation must be good, with no condensation.

PERMANENT VS MOVEABLE HOUSING

Perhaps the main possible negative aspect of any permanent building is the fact that, because a pig cannot go down to the end of the garden in order to defecate, you will, or most certainly should, be removing their droppings as part of the daily, not weekly, routine. The advantages of the ark and paddock system is that there is, primarily, no mucking out. Another is that pigs are happiest when they can dig and forage.

Breeding and veterinary-wise, permanent housing is possibly a plus: should you unfortunately have an animal that requires medical attention, a completely outdoor system might make life more difficult than if it were it housed in a solid building – even more so if the paddocks are remote from roadways. Add to this that a sow about to give birth will usually object to the presence of other animals and you may therefore need to isolate her in another ark, the options of a series of sties or horse-type loose boxes may be preferable.

LEFT Traditional arks give shelter and the opportunity for some free-range foraging.

INSET LEFT Warm comfortable housing results in contented pigs.

TWENTY-FIRST CENTURY HOUSING ALTERNATIVES

Although many of the housing opportunities available for pigs are based on traditional tried-and-tested principles, pig housing is gradually coming of age. The adverts for many arks show that, while the general shape has remained the same, the materials being used are evolving.

The ends of a conventional pig ark are almost always made of timber (usually heavy-duty plywood) – these are now being replaced by polymer, which is not only hygienic but also rot resistant. Those that continue to supply arks with traditional ends are using much thicker exterior grade ply (taken from renewable sources) and some even provide the option of self-assembly – a sort of flat-pack wardrobe but hopefully without the vital bit missing!

There are even 'eco' arks made entirely from farm plastic waste, and bearing in mind just how much of it resides in even the tidiest of farms and smallholdings, this must surely be the ultimate in recycling? Animal Arks (www.animalarks.co.uk), for example, manufactures eco pig arks in a range of sizes, suitable for varying numbers, ages and breeds of pigs. These are of a heavy construction with 12mm (½in) thick walls with industrial-grade fixings to give years of trouble-free use. The roof panel is even available in a choice of colours. In addition, they are said to be rot-proof, easy to clean and simple to move from place to place, by means of the fact that the arks are equipped with lifting eyelets. As might be expected, such arks cost a little more than others, but what price the environment?

Another relatively recent innovation is the fact that many arks are supplied with the option of a floor – previously, the shed would simply cover a patch of earth and litter such as straw would form the basis – as a result, pigs can rest comfortably without their body heat drawing up the damp from the ground. In addition, a solid artificial floor will retain the animal's body heat far better than impacted earth making the ark far cosier for its inhabitants.

BELOW 'Eco' pig arks made from recycled waste farm plastic are the future of pig housing.

DIY BUILDING

'Ramshackle' and 'bodge' is not for pigs; they will investigate every possible loophole in your system and exploit it. Therefore, start with a firm plan and follow it through. Should, for example, you be considering the construction of at traditional sty, make your cement block work or brickwork substantial, as pigs love to scratch and rub against walls. Take the advice of an experienced builder whenever you can as he will be able to advise on the cement mix strength needed and perhaps even recommend the number of blocks required.

Generally, when using standard cement blocks as walls in an existing large building such as a barn, they will rarely need to be more than five blocks high, as, despite the fact that some breeds love to look over a wall and converse with each other in much the same way as cartoonists depict gossipy women over a garden fence, pigs are not good climbers and would generally only ever attempt to escape if they can go under or through an obstacle – and then only if they can see daylight.

ADAPTING EXISTING BUILDINGS

Provided you are able to adjust the floor for drainage and also ensure plenty of natural light and ventilation, an existing building can be easily adapted. Disused stables can be ideal, they should already have drainage and solid floors and two or more pigs can easily be accommodated in the space vacated by a horse. However, horses are taller than pigs and you may well have too much headroom; consider therefore fitting a couple of timbers over their sleeping area, well above

TOP Adhere to the old adage 'measure twice and cut once' if you wish to avoid expensive mistakes and wastage.

LEFT It is important to ensure that all joints and fixings are solidly secured.

ABOVE When constructing an ark any joints and fixings should be made as firmly as possible – the subsequent inhabitants will soon find any weaknesses in your DIY skills.

their reach, and using this shelf for the storage of spare bedding. This will reduce the headroom and also help keep the occupants warmer in the winter.

DOOR REINFORCEMENT

Whether you build from scratch or adapt what you already have, the doorway is probably the most important aspect: pigs are very intelligent and they know that a door opens – they may not know how but they know that it does and will always investigate and attempt to open it (especially if dinner is late). Over time, even through generations of pigs, a poorly constructed door will fail, usually when you are not around – and this is experience talking! As the block work progresses, measure carefully where the bolts will come and insert into the blocks a short length of steel pipe (perhaps about twice the length of your middle finger). It should be of a slightly larger diameter internally than the locking bolt that you intend to incorporate – this is to permit the bolt to slide into a solid, virtually indestructible hole. Without such precautions, frequent rattling by the inmates will eventually fracture the cement. Once again, experience suggests that a bolt fitted onto a surface-mounted metal plate will eventually fail.

The (hopefully substantial) gate hinges must be bolted to the cement blocks with expanding bolts rawl-plugged into holes drilled specifically for the purpose. After everything has been constructed and appears to be correct, fit a substantial 'stop' inside the door at the bolt side, made from a length of timber attached to the block or brickwork – this will prevent the gate being inadvertently pushed inwards and damaging the hinge bolts, as well as reducing the pigs' attempts to rattle the gate. The bolts should now slide tightly; if they don't, pack out the timber stop a little – it will do little to contribute to the strength, but will certainly give the owner greater peace of mind!

ARK CONSTRUCTION

When building an ark, from the ground up, you need: a pair of skids or runners to permit the thing to be moved from place to place. Next, a solid timber floor – many field arks have none but the pig is happier off the bare soil; it retains body heat better and does not transmit damp. Discarded tongue-and-groove floor boarding can be utilized, nailed firmly to cross bearers and the nails punched in, as you would your own household floor. Thick plywood sheeting can be used, but may need extra support. Treat the underside of any flooring with a proprietary preservative as you build and, provided the floor is not in contact with the soil, it should last a lifetime.

As for the roof, the traditional curved shape is the best because it leaves no corners. If you are using corrugated iron for the outside, it will curve relatively easy if laid lengthwise. To prevent the extremes of cold and heat and also to act as insulation, a simple idea might be to insulate the roof, first with chicken wire, then with a layer of straw, before nailing on the sheeting with the special long nails supplied for the purpose. The ends of the ark should be stout and, depending on whether wood or polymer is chosen, either nailed or bolted to bearers. As a last thought, consider fitting a small 'hatch' at the closed end, as when a farrowing sow needs help, she is usually lying with her head towards the door and you may otherwise have difficulty getting past her.

STRAW-BALE HOUSES

This section would be incomplete without the mention of temporary accommodation. As previously stated, the pig will soon destroy anything remotely flimsy, but a friend of one of the authors built a temporary home that lasted six years! It was constructed of the standard small straw bales, each bale was secured by a strong hazel stake driven through to the one below, and the interior was lined with chicken wire similarly pegged. The roof consisted of a number of straight woodland branches topped with another layer of bales of straw. It was, even in the middle of winter, the warmest place ever known and was only discarded when the roof began to fall apart. The cost would have been minimal and could be considered if you have easy access to clean, small bales. You could even consider them in the paddock system and put one in each paddock – very environmentally friendly and not at all unsightly.

BELOW A simple frame and a few straw bales will provide an adequate temporary shelter for hardy breeds but will soon start to show signs of wear and tear.

ARTIFICIAL LIGHTING

No matter what livestock is being kept (and even if the building is to be used only as storage), some form of lighting is always handy and frequently essential. If your chosen method of pig keeping involves a small ark tucked well away in the orchard, while it will be ideal for your pigs, it might not be so good for you during the winter when it may well be necessary to see to your animals in the dark, either before or after work.

In such a situation, you might have no alternative but to don one of the very useful torches that is attached by a strap to your forehead. If, however, the distance to lay an electric cable (either under or over ground) is not too far, then it would well be worth considering including some sort of electric flood lighting. If you decide to supply power via an extension from your home or from another building, installation should be carried out only by a professional electrician. Not only is there the obvious danger aspect to consider, but also the fact that although some DIY electrical work is allowed, it must always comply with the requirements of relevant building regulations.

Where a light is capable of being attached inside the pig ark or part of a larger building where they sleep, it is worth remembering that pigs, being inquisitive creatures, can stretch – often by standing on their hind legs – and thus reach the most unexpected heights. Therefore, it is important to ensure that lighting units and bulbs are always covered with a secure wire cage. Some lighting units are supplied with them as a matter of course, but if not, it is possible to buy one at most agricultural or electrical suppliers.

Finally, it may be that, in some situations where your buildings are too far away from the house or a convenient electricity source, you have no alternative but to incorporate the price of a generator into your initial budget costs. Bearing in mind the outlay, it's important to buy a generator that's suitable for the job you have in mind. A petrol one is generally cheaper and quieter, while diesel has the advantage of being economical in prolonged use, and most diesel models have a larger fuel tank capacity than petrol ones.

ABOVE Any electricals that are accessible must be made completely animal-proof.

BELOW If your buildings are too far from an electricity source, a generator is the only option.

TYPES OF BEDDING

Arguably the best bedding for pigs will be freshly baled wheat straw. The difficulty is that many farmers now bale their straw in large round or square bales, which will be difficult for you to handle or store. Some farmers however, recognize that there is a ready market from smallholders and stables, and bale straw specifically for this trade.

Wheat straw should be baled as fresh as possible, ideally following the combine harvester. Once the straw has become damp, even from heavy dew and even though it will dry out, spores that are not good for the pig's respiratory system will have begun to grow. Look at a sample of the straw before you buy it, it should appear bright, almost polished and have no noticeable dust of dark mould emanating when you shake a handful.

If you find such good straw, the price suits you and you have room to store it, purchase as much as you can – if you can collect it direct from the field you may get a better price and certainly a fresher product.

Barley straw would be the next choice, once again freshly baled. The problem is that barley straw contains the 'awns' or beard seen on the seed head and these can cause itching, not only for the pig but also for you when handling it. However, do not let this dissuade you if you have little choice.

Of wood shavings, beware! Certain woods give off dust that is carcinogenic and must at all times be handled wearing facemasks or respirators. Pine is the best choice; you may learn of a factory dealing only in pine furniture and be able to secure a regular supply. If their dust extraction system is a modern one it will separate the shavings from the fine dust that can cause problems. The other problem with using shavings will be disposal, as it is unfit for compost even when saturated with urine and mixed with faeces and will need to be piled up somewhere for a number of years before it can be used on your garden – otherwise it will de-nitrify your soil while it is breaking down.

Finally, consider shredded paper: many office blocks have a large quantity of paper waste from their shredders and may even give it to you to take away. However, you may need quite a lot of it. More practically, bales can be bought from your agricultural suppliers.

LEFT Many farmers now bale their straw in large round bales rather than the smaller ones more useful to the pig-keeper.

ABOVE Wood shavings should be clear of dust.

PLANNING PIG PADDOCKS

Before you go to the expense of fencing your paddocks, it is a wise move to begin with a simple paper plan. The first consideration ought to be whether it can be practically divided – it is far better to have at least two paddocks in order that the ground can rest; three or four will be even better, but it rather depends on the overall size of the plot.

For the sake of your paddocks as well as the health of your pigs, four small plots are better than two large ones: if you keep pigs on a small patch of land, without good care and a regular rotation, they will ruin it and it will be extremely hard work to bring it back into a good state of repair.

It is important to consider the terrain, the slope and the aspect; a paddock consisting of just a low-lying bog that rarely sees the sun is, for example, better incorporated with another piece of land that does not have this drawback. If there are wooded or scrubland type areas, try to divide the paddocks so as to allow a little in each one. Shape is unimportant, content is.

Finally, on very rough land, determine a good clean line for each fence, negotiating a way round large rocky outcrops rather than attempting to go over them, as to try and do so will make any kind of fencing difficult. If you end up with three or four paddocks, each with some clear ground and a portion of scrubby woodland, you have 'pig paradise' and are fortunate indeed.

ACCESS

You will need to be able to get a vehicle to the paddock from time to time in order to deliver arks, to transport animals and food, and your veterinary surgeon will definitely appreciate not having to trudge over muddy fields to offer assistance as and when it will, unfortunately but undeniably, be required. From your own point of view, no matter how enthusiastic a pig-keeper you may be, the novelty of an off-road trek across muddy land will soon wear thin as winter progresses – even the simplest of roadways made up of coarse gravel will help, and can be added to over time. Most farm tracks began in such a way. Gates are expensive – both to purchase and to fit – but it is still better to consider permanent ones. The gateway needs to be large enough to permit a tractor through pulling your ark or arks. A saving can be made by having just one main gate to the first paddock, then a

lesser one into each of the others, but even so it is important to remember that they should be wide enough to be able to pull the ark through from one to the other.

For personal day-to-day access, the cheapest alternative might well be a simple homemade stile from one paddock to the other, but you should prevent your pigs from being able to get near them so that they are not tempted to use them as scratching posts. Alternatively, if you can manage the cost, a simple small pedestrian gateway might be considered.

SERVICES

In the planning stages, you should also take into account the possible provision of services such as electricity and water. Assuming the distance is not too great, consult an electrician before you begin – he or she must be qualified to install the electrics and will advise on necessary cables and fittings. If, on the other hand, the intended paddocks are too far from home to make an electrical supply a feasible option, you will either have to consider car batteries, a generator or doing without!

The absence of water is a far more serious problem and if you are obliged to use a portable trailed bowser, you will need to consider carefully how you intend to secure it, both against theft and tampering. For either electricity or water you can save on costs by digging the required trenches yourself, but remember that the depth needs to be sufficient to ensure the water supply will not freeze in winter, or be accidentally unearthed by the rooting of your pigs should it run through one of the paddocks.

OPPOSITE Scarifying and periodically resting paddocks will help keep them in tip-top condition.

ABOVE Where finances allow, surround any temporary paddocks with a perimeter of permanent stock-proof fencing (see background).

FENCING YOUR PADDOCKS

Electric fencing is cheaper than permanent fencing but needs frequent observation to ensure that all is working well and that you avoid the possibility of a pig throwing soil or turf over the wire and shorting out the circuit. Being the intelligent creatures they are, it won't be long before your animals discover it is no longer live and make a bid for freedom. A permanent fence will have a greater initial cost but once erected should last for 15–20 years with little maintenance.

Treated timber posts should be either driven or dug and cemented in every 5m (5yd). At every major change in direction, corners, gateways and every ten posts, a more substantial post called a strainer needs to be installed. Three plain galvanized wires should be added next, tightened and fixed to the post at the top and bottom of the stock fencing and in line with the middle wire of the fencing. They need to be fixed to the posts with galvanized fencing staples but should never be driven in fully; this permits the wire to move with expansion and contraction. Were they to be fully driven home, there is a good chance that the coating will abrade and rusting will occur – eventually leading to the wire breaking at that point. Not hammering them right into the stakes also makes it easier to remove them when it comes to replacing an accidentally broken post.

The strainer posts and corner posts should be braced with an angled post in line of the fence and notched to the post before being secured with a spike or large nail. Finally, the stock fence should be 'offered up' to the strainer wires and here it is important to note that there is only one way to erect stock fencing and that the smaller squares must be at the bottom, their design and purpose being to keep in smaller stock such as piglets. The stock fence should then be pulled tight and secured to the strainer wires with what are appropriately enough known as 'pig-tails' – pre-twisted wire coils. At gateways, the wire needs to be finished by wrapping it around the gatepost tidily and securing it back to itself with staples that, once again, are merely to hold the wire in position and should never be driven fully home.

BELOW Upright posts should be quite close together and the wire attached to the inside rather than the outside.

FENCING THROUGH BOG

If it is unavoidable that your fence is forced to cross a boggy piece of ground, make certain that it is not possible for pigs to make themselves a nice mud bath and then discover they can dig out! Avoid the possibility by cutting a trench along the fence line until you are into firmer ground (use longer posts if necessary) and then, along the bottom of your trench, add an extra piece of straining wire before cutting a length of stock fencing and attaching it to the extra straining wire and the normal bottom strainer wire. Then, when all is complete, bury it! Better now than when a precious animal has absconded and remedial work is called for as a result.

GATES

Gates may be expensive, but buy the best you can afford (as you should with all your fencing materials), ensure that any label attached carries proof of its quality – such as the American SEI quality mark or the European CE conformance mark, for example – and you will not be disappointed The label should refer to the strength and construction as well as the thickness of the coating or galvanizing. There are a lot of cheaper gates on the market, mainly originating from factories in China or Eastern Europe and you may be tempted to make savings. Be warned – they do not comply with standards, may be of very inferior construction and it could well be a false economy.

The best gates for pigs are of tubular coated steel and have a mesh panel right along the bottom section – this serves to keep piglets in, dogs out and deters children from swinging on them! They should have good lockable bolts and, as extra security, fit the bottom hinge, 'pin up' and the top hinge 'pin down' – while this may be a little tricky, it means that should anyone attempt to steal your gates, they will have to dismantle the hinges completely.

ABOVE Pigs are inquisitive creatures and will soon discover any weak points in your fencing.

RIGHT No matter what type of gates you choose, they should always be well constructed.

ELECTRIC FENCING

The electric fence is ideal for large areas where permanent fencing would be prohibitively expensive. Areas can be fenced for a period, then, when the animals have made the best use of the ground, they and the fencing can be simply moved. This system, known as 'folding', has also been used for generations with sheep.

The electric fencer is, in essence, a small 'box of tricks' powered either by batteries or a line from a mains supply. The unit or energizer gives out a powerful electric shock every few seconds, which causes no harm to animals or humans, but neither enjoy the experience and, after a few contacts, both will learn to avoid it. A number of small plastic or metal posts are pushed into the soil at distances around the area you wish to 'enfold'. These are fitted with moveable insulators capable of being adjusted for height and may be as few as one for a single line fence or three or four for something more substantial. The wire or wires are then attached to the insulators and pulled taut to avoid sagging (always use single strand, never twisted, and consider using tape rather than wire as it is easier for pigs to see). Finally the fencer unit is connected and switched on.

As long as none of the wire strands of the fence are touching soil or vegetation, the pulse of electricity will keep animals inside the temporary paddock. Remember to test once or twice a day to ensure all is well: if it is not, check the battery or feed and if that's OK, walk the line until you find the cause of the 'earth'. The wire may have sagged, a branch may have blown onto it, or quite often an enthusiastic pig will dig a large turf and it will fall against the wire.

One final important consideration with pigs is never to use electric fencing across a gateway. They will remember it and even if you dismantle it completely when they have to be moved, they will be very reluctant to cross where it has been. Better to install a small but secure gate and terminate the live wire each side of it.

BELOW Electric fencing is cheap and practical but must be checked regularly to ensure it is not 'shorting out'.

'FOR THE WANT OF A NAIL ...'

Pigs bash, crash and bluster – that is part of their charm. They always feel the need to explore and, being inquisitive creatures, there is generally something of far more interest in pastures new: hence their propensity to quite literally poke their snouts into anything unusual.

ABOVE Make any necessary repairs as soon as you notice them – and not after your pigs have!

They will, given the opportunity, and through no maliciousness, lift a less-than-perfect stock fence (hence the reason for suggesting that you might consider attaching a length of barbed wire underneath any perimeter fencing – see chapter 1) and, with their powerful shoulders, are able to lift not only loose wire, but also any less-than-stout posts from their base.

Periodically then, and as part of your general maintenance and repairs programme, it is a good idea to check both the posts and the staples that attach the wire to the posts on a regular basis. In fact, it might be as well to include a patrol of the fencing as a part of your evening routine (when you will not be as worried about being late for work) and have to hand a hammer and pocketful of spare staples (perhaps stored conveniently in the food shed?) with which to undertake any necessary remedial work.

Pigs also – surely out of curiosity – have a habit of sticking their noses under the corner of a ship-lapped wooden ark and thereby inadvertently loosening any nails. Replace them as soon as the problem is noticed, otherwise it will not be long before half the cladding has gone! As a matter of routine, make sure that hinges and door bolts are periodically oiled and that wear and tear on any hosepipes is remedied as soon as it is noticed rather than allowing a harmless trickle to become a damaging deluge.

As well as the daily and weekly check-ups, there are a few jobs that are specific to a particular season of the year. Traditional wooden houses need an annual treatment of preservative, which must, of course, be animal-friendly. Such care will undoubtedly prolong the life of any pig ark and might save a fortune in avoiding having to contemplate possible replacements in the future.

FEEDING

GIVEN A CHOICE, PIGS WOULD PROBABLY PREFER TO EAT LITTLE AND OFTEN THROUGHOUT THE DAY, WHICH IS WHAT THEIR ANCESTORS WOULD HAVE DONE AS THEY RUMMAGED ABOUT THE WOODLANDS AND HEATHS IN SEARCH OF FOOD. THIS IS, HOWEVER, NOT AN OPTION OPEN TO MANY AND SO MOST PIG-KEEPERS FEED THEIR ADULT STOCK TWICE DAILY. FEEDING YOUR PIGS IS ARGUABLY ONE OF THE GREATEST PLEASURES OF THE WHOLE BUSINESS, AS IT'S GOOD TO OBSERVE THE DELIGHT WITH WHICH YOUR ANIMALS GREET YOU (AND THE FOOD BUCKET!). IT IS ALSO AN EXCELLENT OPPORTUNITY TO CAST AN EYE OVER THEM AND CHECK FOR ANY POTENTIAL HEALTH ISSUES.

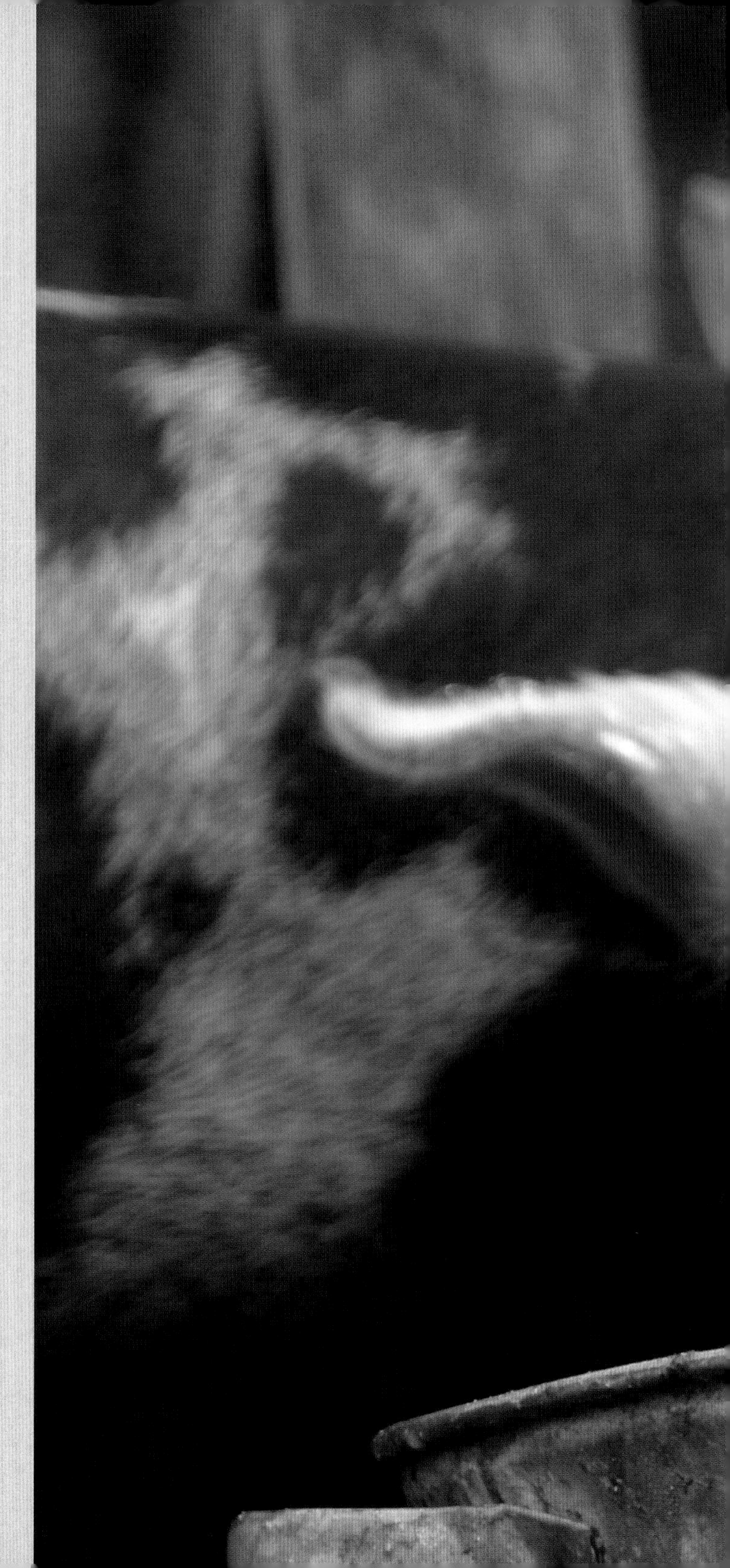

FOOD FOR THOUGHT

One of the traditional fascinations of keeping pigs is the fact that they will eat virtually anything – and so they might, given the opportunity, but you also need to consider the legalities and logistics of feeding. On the plus side, anything that goes through their bodies can be usefully added to the compost heap and subsequently the vegetable garden. Hardened stalks of cabbage and the like, the leaves of which have been eaten, will be 'mulched' by the trampling of pigs' feet and these too can eventually be added to the compost bin.

In order to set the record straight immediately, it is imperative to mention the fact that, in most 'modern' countries, it is illegal to feed catering waste or animal by-products to any farmed animal or, 'any other ruminant animal, pig or poultry.' 'Waste' food is defined as that which comes from animals – including cooked table scraps – or any other food that has been in contact with meat or eggs, including cooking oils. To do so, you are, in the eyes of most government edicts, not only risking your pigs' health, but also breaking the law and could be prosecuted in a criminal court.

Feeding a pig supplements such as stale bread, for example, or acorns and nuts for which you have foraged at the appropriate times of the year is of course, quite permissible (though care should be taken not to feed too many acorns at any one time as the shells' high tannin content could cause problems) – as is the fresh produce of your vegetable garden. Technically, the latter should perhaps not have previously entered the kitchen, as, while it is acceptable to throw the outer trimmings of a lettuce or cabbage to your pigs en route to the kitchen, once it they have been cut on a chopping board, they could have become contaminated by being on the same board that had previously been used to cut up meat. But, while it is a point to bear in mind, it is possibly one step too far for the average hobby pig-keeper!

With many of the old breeds now enjoying something of a revival in popularity, it pays to take advantage of their availability – not least because there is often not the necessity to feed expensive concentrates exclusively, as would be the case with commercially produced animals. However, whatever your chosen breed, never be tempted into overfeeding: it costs you more, the pigs do not benefit and, irrespective of whether or not you are contemplating meat production, you will end up with fatty animals.

BELOW Ensure that there is sufficient trough space for the amount of pigs – if different breeds are kept together there is a possibility that smaller types might be pushed out or bullied.

WHAT PIGS EAT

Pigs, as we have established, will eat virtually anything – and give the very definite impression of enjoying it too! Nutritionally, pigs require a mixture of water, carbohydrates, fats, proteins, minerals and vitamins – most, if not all, of which will be found in a combined diet of outdoor grazing and complementary pellets (pig nuts) or meal.

It is, of course, possible to write a scientific dissertation regarding the exact nutritional requirements of a pig. Suffice to say that, in times gone by, pigs would have been fed whatever was available. Nowadays there are commercially produced feeds that will ensure your animals are given the best possible nutritional combination.

Easy though pellets and meal are to supply, pigs, like humans, nevertheless prefer a little variety in matters gastronomic. In the wild, pigs would quite literally bulldoze through the undergrowth in order to gain whatever they needed. In the spring they might uproot newly emerging shoots and wild garlic bulbs, and nibble the leaves of a tree's lower branches. In the summer even more would become available, while in the autumn and winter, they could live off fallen nuts and fruits and, if times were really hard, gain some additional nutritional value from seeking out hibernating rodents, insects, worms, the decaying bark from trees and even carrion. Today's domesticated pig breeds like to do much the same, so you should offer them every available opportunity to do so alongside a daily handful or two of concentrates.

If space permits, it might just be possible to let your pigs find almost all their food for themselves. Wherever pigs have ready access to plenty of grazing, they will, surprisingly enough, get a good proportion of their necessary sustenance from eating grass. A natural diet will, however, take longer to produce a meat carcass, as such methods are not necessarily as efficient as specialist pig foods and the pigs will also use up energy snuffling about and in keeping warm.

ABOVE When kept in large units, growers are sometimes fed ad lib via hoppers.

FEEDING ROOTS AND GREENSTUFFS

Should you be a vegetable grower then nothing need go to waste when you have pigs. If land allows, it is not a bad idea to grow extra crops in order to be able to feed them really fresh greenstuffs like lettuce, or root vegetables such as potatoes (which are preferred cooked), turnips, carrots and parsnips – all of which will keep them amused and also add natural vitamins to their diet. Never overlook the value of the pig as a potential consumer of your surplus produce – small or 'chat' potatoes, cabbage leaves, overripe tomatoes and much more – pigs are arguably the best composting machine yet invented!

ZERO GRAZING AND STRIP GRAZING

Should you contemplate growing food especially for them then there are two main systems that can be utilized, the first of which involves growing crops and taking the produce to the pig – a system known in farming parlance as 'zero grazing'.

The second method is simply to sow a thin mixture of some or all of the above and let nature take its course. Then, when there is sufficient growth, use an electric fencer, or even moveable temporary fencing, to let your animals have access a small section at a time – a method that farmers term 'strip grazing'. Finally, when the crop seems exhausted, take them away, leave it for another week or so if the weather is mild and then put them back in the whole piece. They will relish any new growth and having exhausted that, continue to dig over the whole plot looking for bits of root. Not only will they have a great time doing it, they will also leave your ground partly dug and well manured.

Your own potato crop will also benefit from their attentions: even after you have dug every potato in sight, any gardener knows that there will be several that have been missed, so why not remove any 'haulms' or potato tops (which are not good for them) before allowing your pigs access? In a very short space of time they will find and consume every last potato, as well as many weed roots and grubs. When they leave, the ground will be in good condition and you won't have the problem of unwanted rogue potatoes next season.

PLANTING AND STORING

For both the above methods, you will need more seed than the rather expensive small packets sold for the garden, so consider contacting an agricultural seed supplier. Here you will be able to purchase in quantities of parts of a kilo and at far more acceptable prices. He or she will even advise you on varieties suitable for growing in your area and timings for planting. With a powered cultivator or small tractor, the task is made easier. Many root crops will remain firm in the soil for several months after maturity; however, you have the option of lifting them in the autumn, trimming off the tops and storing them in a small clamp, covered with a deep layer of straw. Leafy crops such as members of the Brassica family can be grown in the same fashion

and cut as required. Remember though, that the nutrient value of leaf crops is not as valuable as root crops and feeding too much may cause digestive problems. Ensure your pigs' diet is balanced with a high-protein supplement.

A word of caution on the feeding of 'waste' garden produce and forage crops; they must of course be given as part of a balanced diet and in any case, not too much at a time (hence the need for strip grazing when allowing pigs free range over what has been planted for their benefit). Having previously stated that their digestive system is similar to our own (see Introduction), never treat the pig as a dustbin. Overripe produce is fine (a definition for which is that you could eat it if you chose to), however, rotten produce is not as good an option and should really find its way straight to the compost heap. Items such as rhubarb leaves, potato tops and citrus peel have no place in a pig's diet and could cause harm. Let common sense be your watchword – if you could eat it, so will a pig; if not, don't feed it.

OPPOSITE Surplus greens from the vegetable garden will be appreciated by your pigs.

BELOW Strip grazing allows for controlled feeding – important factors when it comes to porcine health and crop wastage.

FEEDING CONCENTRATES

Pig concentrates are predominantly made up of cereals and vegetable proteins such as soya bean, and can be made up as either a meal (mash) or pellets, sometimes known as pig nuts. There are many pig-keepers who prefer to give the latter, not least because they are cleaner and less dusty to feed.

Pellets come in varying sizes – ranging from a weaner pellet, which might be around 3mm (⅛in) in diameter, to ones around four times that diameter for adult stock. In addition, many manufacturers supply an option of 'rolls' that are designed to be fed on the floor or scattered into straw – making them ideal for hobby herds of outdoor-based pigs. The type of concentrate required will, of course, depend on their stage of growth and subsequent nutritional requirements. Fortunately, any sort of balanced diet is readily available from your local agricultural suppliers and include those most suited to breeding pigs, pregnant sows, lactating mothers, weaners, growers and for general maintenance.

With certain exceptions, pigs should generally be fed twice daily rather than having concentrates available all the time (and here the word 'concentrates' must be emphasized because there is nothing wrong with having natural supplements such as greenstuffs on hand throughout the day) and a regular, timed feed is very important to all pigs, no matter what their stage of growth. Leaving food in the trough is a waste, as it can be eaten by birds and vermin, and can also go mouldy if left for any amount of time, which is, incidentally, another reason to clean troughs and feeders on a regular basis. Just as important is the fact that with too much food and too little exercise, pigs will rapidly become overweight.

BREEDING AND LACTATING SOWS

When kept for breeding, pigs need to be fed a little more than the average general ration. Pregnant sows need food that contains plenty of protein. Manufacturers generally have their own specific formula (and name) for the various concentrates given at different stages of a pig's life. The ones quoted in the following short section are typical, and are given simply as examples of those produced by many feed mills.

Breeding sows (and boars) are fed on 'sow rolls' or 'cubes' that consist of a good cereal base and around 16 per cent protein. They might then be fed on 'super sow' cubes or rolls during lactation. The latter is higher in protein (at around 18 per cent) and includes most if not all of the necessary vitamins and minerals that a sow with youngsters should need.

Make sure that the sow cannot have access to any of her piglet's weaner or grower food; not only will it cost you a fortune, there is a very real danger that, because of the extremely high protein levels they contain, she may well suffer from loose or even watery stools.

WEANERS AND GROWERS

Serious attempts at weaning can begin when the piglets are about three weeks of age, but you should nevertheless begin putting milk pellets in a shallow trough in the creep area (see chapter 6) from the age of anywhere between seven to ten days. Begin with approximately two heaped teaspoons per piglet, and although they will eat next to nothing of them for a while, the amount can gradually be increased as the piglets become more accustomed to their taste and texture. Some say that the food at this stage should be fed ad lib, while others advocate removing whatever hasn't been eaten within an hour. Typically, weaning pellets will contain around 20 per cent protein, which should then drop to about 17–18 per cent by the age of 12 weeks.

If anything, young pigs are even more inquisitive than adults and will taste and chew away at whatever takes their fancy. It is therefore vital that nothing is left around that could do them harm – and this includes shards of broken glass, sachets of vermin poison that may have been dragged out from under a less-than-secure baiting point, and the ubiquitous strands of baler twine that seem to be found around every property where livestock is kept.

ABOVE In order for a lactating sow to produce a plentiful supply of milk, she needs to be fed concentrates with high protein content.

OPPOSITE Pig nuts are easy to source and contain all the necessary nutrients.

HOW MUCH TO FEED?

As with any animal, the daily amount of food rations necessary varies with the size of the pig and whether or not it is able to pick up any supplementary food from the paddock where it is kept. The simple and somewhat glib answer to the question of how much to feed when it is asked of an experienced pig-keeper is sometimes given as 'feed to appetite', but that is often easier said than done.

Basically, this means that if your pig is too fat, too thin, ravenously hungry or, conversely, not hungry at all, you should feed accordingly. Very generally, however, piglets at the weaning stage should be eating approximately 350g (12oz) per day, which rapidly rises to just under 1kg (2¼lb) at the age of two months; six-month-old pigs need 2.35kg (5lb); outdoor-kept pregnant or lactating sows will consume up to 3kg (6½lb) per day (but almost twice that if they are kept intensively) and an adult boar will chomp his way through about 2.65kg (5¾lb) during the course of his day. Of course, manufacturers have developed many and varied pig mashes and pellets so you should be guided by their recommendations – either displayed on the bag or obtained by ringing their advice line (the number for which will almost certainly be printed on the bag).

As far as meat production is concerned, it is reckoned that feeding pellets rather than meal will give a better food-to-meat conversion rate (approximately 7.6 per cent better, in fact). Depending on the breed, fattening pigs could possibly be given food ad lib until they are about half-grown, after which time it should be restricted to what they can eat during a twice-daily feeding.

In all other circumstances, it is usually recommended that pigs be fed twice daily and only given as much as they can consume in say, a 15- to 20-minute period (this restriction helps in preventing them from putting on too much surplus fat).

Rather than feeding your pigs and then rushing off to see to something else that will undoubtedly need doing as part of the daily routine, it is better to get into the old stockman's trick of leaning on the gate and watching them eat – it is at such times that you will be best able to monitor the health of your pigs and, if several are kept, ensure that none of them are being bullied and not getting their share of the food.

ABOVE Many pig-keepers opine that pigs should be fed only as much as they can eat in a 15- to 20-minute period.

VITAMIN AND MINERAL SUPPLEMENTS

Pigs need a supply of minerals and vitamins – a good proportion of which they can often pick up as a result of rooting about in their paddocks and runs; they are also included in the concentrates bought from your agricultural supplier. It might however, be useful to know the exact purpose of some of these minerals and vitamins in order to decide when, and if, any additional supplements might need to be given.

Calcium and phosphorus are essential for all stages of growth, gestation and lactation – limestone is an excellent source of calcium and will be ingested naturally in places where it forms part of the soil structure. A calcium supplement may need to be given to lactating sows and young piglets.

Sodium, chloride, potassium and magnesium will be found at the correct level in any proprietary pig food and no supplements should be required (although at one time it was thought that a magnesium supplement could help prevent cannibalism in commercial herds, research has since proved the theory to be erroneous). Because the amount of iron in sow's milk is very low, it could just be necessary to provide supplemental iron in the form of iron dextrin, often given as an injection, during the early stages of a piglet's life. Manganese may be required by the sow during pregnancy and also while feeding her litter, while zinc has been found to be beneficial in reducing the possibility of piglets having post-weaning diarrhoea.

Vitamins A and E are essential for many things, but in addition to those found in compound foods, all that a healthy pig is likely to require can be discovered in green foodstuffs, which it will find free range, or in vegetables thrown into the pen. Outdoor pigs will get some of their necessary vitamin D from the sun's rays, but don't forget to offer shelter and a wallow so that they don't get sunburnt. It is possible to buy multivitamins for pigs, but before dosing up your animals with any of these, it is essential first to seek the advice of your veterinary surgeon or knowledgeable feed supplier.

BELOW Pigs can sometimes be given supplements or medication by drenching. The drench can be given from a bottle but must be carried out carefully to ensure that no liquid enters the lungs.

LIQUID ESSENTIALS

One half to two-thirds of a pig's body is made up of water and they should be given access to fresh, clean, cool water at all times. Water also plays a vital role in regulating their body temperature, food digestion and in expelling waste body matter.

The exact amount of water a pig needs obviously varies with its age and the type of food it is being given – an animal fed only on concentrates will drink much more than one that takes in a certain amount of natural moisture as a result of eating greenstuffs and root vegetables. Lactating sows consume more water because of the high water content of the milk that they produce for their piglets. The ground on which a pig is grazing might also have some bearing on the matter – pigs kept on estuary land where the natural growth contains a high proportion of sea salt, for example, will undoubtedly drink more on a daily basis than their inshore counterparts. As a very general guide, piglets will drink two litres (3½ pints) of water a day, dry sows 12 litres (21 pints), suckling sows 20–30 litres (35–53 pints), and a boar 10–20 litres (18–35 pints).

Although you will no doubt on occasions see your pigs drinking from puddles, such practice should be avoided wherever possible, as the inevitable bacteria found there could have a detrimental effect on their health. For the same reason, it is best not to use rainwater collected via guttering and drainpipes and stored in barrels. Suitable though such water may be for watering the vegetable garden, you can never be sure exactly what impurities it contains. Likewise, be careful about using well water, which might contain naturally occurring, but nevertheless potentially damaging levels of sodium, chloride sulphates, magnesium and copper. If you read the contents of your food bag carefully, you will find that all of these minerals are included – but at a veterinary prescribed dose. Include a water supply that contains even more of the same and your pigs may well suffer.

BELOW Water must be available at all times. Trough water can rapidly become dirty and full of straw and should be changed regularly.

FEEDING AND DRINKING TROUGHS

In dry weather, many aficionados like to throw pig nuts on the ground – which keeps the pigs far more amused and busy than if they were fed in a conventional trough. One of the most popular feeders best suited for pigs is the round 'Mexican hat' cast-iron trough, which is excellent for feeding litters or smaller pigs. They are however, difficult to move – a plus point when it comes to boisterous pigs being unable to push it all around their paddock, but not so whenever you need to move it to another place.

In the interests of hygiene, one thing that is important is the need for feeders and drinkers to be easily cleaned and periodically sterilized: the most sensible way to achieve this is to take them away to wash them. Obviously this is not possible with automatic drinkers, but it should still be a matter of priority to see that they are kept free of errant food particles. Galvanized or alloy utensils are more robust than even heavy-duty polyurethane but they are, unfortunately, often as much as twice the price. Both will clean just as easily although it has to be said that the polyurethane probably has the edge, as they are usually constructed in a mould, whereas metal-made feeders and drinkers have seams and lips that are sometimes difficult to scrub. Under no circumstances be tempted into making your own feeders out of wood – they will be extremely hard to clean thoroughly and there is also the very real chance that second-hand timber may have been treated with paint or preservative that is harmful to livestock.

When it comes to automatic drinkers, although there is undoubtedly a greater convenience in having such things attached to the mains water supply or a water bowser, some experienced pig-keepers don't particularly like them because they feel that a pig likes to 'drink deep' and is unable to do so in most types of snout-operated automatic trough. They also mention the fact that in hard water areas, there is a potential problem in that there could be a gradual build-up of limescale in the pipes – which will obviously restrict the flow of water getting to the pigs. Should you have an automatic water trough that is deep-sided and to which piglets have access, a couple of bricks or cement blocks placed in the bottom will assist them in scrambling out if they happen to fall in.

ABOVE Pigs like to drink deep, so many pig-keepers prefer not to use automatic troughs.

STORE SHEDS AND STORING FOOD

Where space precludes the provision of a specific building or store, at the very least you should keep food in vermin-proof storage bins. You can use galvanized dustbins (don't be tempted into using plastic, as rats and squirrels will soon make short work of gnawing through the most vulnerable points), but far better are the sloping-lidded food bins much favoured by horsemen and poultry keepers of bygone eras. To buy new these would be extremely expensive but it is sometimes possible to pick one up reasonably cheaply at a farm sale.

In terms of the positioning and type of building most suitable as a food store, unless you are prepared to waste produce and money, a leaking lean-to will not do. Not only is there a chance that the food being stored there will be spoiled, it is also highly likely to get damp and grow mould, which will be harmful to your pigs. The shed should be dry and, crucially, totally vermin proof. You can help in ensuring against the latter by keeping a regular check on likely entrances and exits such as the bottom of doors, places where water pipes go through the walls and under the eaves of a building that has vegetation growing up the sides. Fill any gaps, place small mesh netting under the eaves, and fit a galvanized strip along the bottom of the door so that rats cannot chew their way in. Most importantly, keep a constant supply of rat poison both in the building and in safe strategic baiting points outside: the type that comes ready prepared in sachets is probably the best, as the plastic the sachets are made from helps to ensure that the bait is kept dry and fresh until such times as a rat or mouse chooses to chew through it (vermin loves to chew through plastic and will naturally eat the bait after doing so). Check all baiting points regularly – infrequent baiting is as useless as doing no baiting at all, and may lead to a build-up of immunity. Make sure that all baits are inaccessible to your stock or indeed to any other domestic animals.

BEST PRACTICE

Inside the shed, all bagged food bought from an agricultural retailer should be stored off the ground – even a concrete floor is never completely dry, so it is imperative to ensure that air can circulate freely. Buying in a large stock of food may seem economical and cost effective, but only if it can be properly stored on wooden pallets and away from the walls so that air can circulate under and around the bags. Always keep a note of what food is stored where and try and to develop a rotation system whereby the first food purchased is the first to be used – it is all too easy to store food next to the door and use it first from this, the

most obviously accessible pile. If you fall into this trap, the older food at the back of the shed will go stale and unpalatable and any preventative drugs, medicines and vitamins will most likely lose at least some of their effectiveness once out of date. Get into the habit of checking use-by dates on purchase and never use food that has gone beyond this time.

If it is a big enough building, a part of the food store can be apportioned for storing straw or other bales of material being used as floor litter (see chapter 4). Again, these should be stacked neatly and tidily – as should any empty food sacks or redundant pieces of baling twine and the like. There is no doubt that these items may well be useful at a later stage (paper sacks for storing foraged nuts and acorns, plastic ones for acting as temporary wind baffles at the window of a farrowing ark, and string … well, however could we manage without it?), but they should be kept in bundles tied up and out of the way in the roof space rather than strewn all over the floor. A cupboard in the store shed is a useful asset; in it can be kept vitamin supplements, medication and sundries such as spare elements for infrared heaters – not only will this help keep everything together, you will have at least some chance of remembering where a certain item might be!

OPPOSITE Food storage bins must be made from galvanized metal to prevent gnawing and rust.

BELOW Whether you have one sack or 100, all food must be stored off the ground so that air can circulate.

BREEDING

IF SPACE PERMITS, THERE WILL ALMOST CERTAINLY COME A TIME WHEN, EITHER FOR FINANCIAL REASONS OR FOR THE PURE PLEASURE OF PERPETUATING THE BLOODLINE OF A PARTICULAR PIG, YOU MAY WISH TO CONSIDER BREEDING A LITTER OF PIGLETS ALL OF YOUR OWN. WHILE IT MIGHT NOT BE AS SIMPLE AS, SAY, PLACING A BATCH OF FERTILE EGGS UNDER A BROODY CHICKEN, IF EVERYTHING IS CAREFULLY THOUGHT OUT WELL IN ADVANCE AND POTENTIAL PROBLEMS NEGATED BEFORE THEY HAPPEN, THERE IS NO REASON AT ALL WHY THE EXPERIENCE SHOULDN'T BE BOTH SUCCESSFUL AND REWARDING.

BREEDING BASICS

Can there be any more attractive a mental picture than a contented sow snuggling down in a bed of fresh straw surrounded by piglets feeding from her ample teats? But before that happy scenario occurs, there is much to contemplate regarding the practicalities.

There may be several reasons why you want to breed from your pigs, not the least of which is the very valid fact that you are helping to perpetuate the genetics of some of the old traditional and/or rare breeds, which without the enthusiasm of dedicated small-scale breeders might otherwise become extinct. If you are interested in making a profit – or at least breaking even – breeding a litter, part of which you can sell on (perhaps keeping a couple back for home use) should help in achieving this happy state of affairs. It might also just be that pigs bought primarily as pets or in order to help clear bracken from woodland, for example, become so much a part of your life that you eventually decide to breed from them.

If you are intending to show or keep very clearly to the bloodline purities of a particular breed, individual pigs will undoubtedly need to be registered with an appropriate association. In the UK all pedigree animals should already have had their details lodged with the British Pig Association (BPA) from a very young age, while in the US there are several such organizations, including the National Swine Registry (NSR). Irrespective of whether it is pure-bred or pedigree (the subtle difference between the two having been explained in chapter 2), it is important to ascertain that the pig you wish to register is either herd-book registered or is otherwise eligible to be used as a breeding animal.

So, how best to start? Generally, if you have had breeding in mind from the very beginning, you might consider building up the bones of your little empire by buying a pregnant sow and then, when you've gained a little experience as a result, subsequently mating her with a nearby boar or, depending on your chosen breed, perhaps even contemplate finding out more about artificial insemination (AI).

BELOW A healthy, lively looking litter of piglets.

WHEN AND HOW?

A young female pig will first show signs of being physically able to breed as early as six months, but it is far better to wait for at least another couple of months before considering doing so. Most people agree that a gilt should not be bred from until she is almost a year old, because by then she should be a strong animal in good body condition, and therefore more able to withstand the undoubted rigours of mating and motherhood.

Generally, a sow will show signs of actually coming on heat (or being 'in season') two or three days before she is actually ready for serving (mating): the most obvious indicator being an engorged dark-red vulva plus mucus. As she becomes more receptive to mating, additional signs may include becoming unusually noisy. One well-known old-timer's trick to test whether a sow is properly on heat is to stroke and gradually push down on her back. As you do, she should stand stock-still and may even prick and twitch her ears more than would normally be expected. If you have other females with her in the herd (which you should because, as has been previously mentioned, pigs should never be kept alone – except of course, when farrowing or perhaps being medicated), they will sometimes try and mount her.

Normally, a sow comes on heat every 21 days, but individuals have been known to range between 15 and 25 days; therefore, if you are finding it difficult to monitor when she is exactly ready, it is possible to interfere with nature and use a hormone progesterone, which will break the cycle and, once it is no longer being given, will result in her coming on heat five days later.

If you are only keeping two or three females, it is probably not economically viable to keep a boar as well and so when the time comes that your sow is ready, you will need to take her to the sire – but choose him carefully as obviously he represents half of the genetic pool. It is therefore essential that you have contacted the owner of a potential mate well beforehand. Try and find an owner who is willing to keep your sow for a few days when she is in season, as otherwise, you may make the journey to the boar only to find that the female isn't really ready and needs to wait another day or two.

ABOVE A thriving herd of healthy breeding pigs.

TAKING THE SOW TO THE BOAR

In a book of 1789, Gilbert White, the famed naturalist and vicar of Selborne, Hampshire, told of how, when a neighbour's sow came into oestrus and wanted to mate with a boar, '... she used to open all the intervening gates, and march, by herself, up to a distant farm where one was kept; and when her purpose was served would return home by the same means.' The more usual course of events is, however, for you to take the sow to the boar!

On arrival, the boar owner should help you in introducing your pig to the male, who, if he is experienced in such matters, will certainly know whether or not your sow is ready and on heat! A natural mating or 'service' can take quarter of an hour to complete – which is a long time for a sow to stand steady with a heavy boar on her back and is one of the reasons why a gilt should be a mature, strong animal before being served for the first time.

Although AI is perhaps more commonly used in commercial pig-breeding establishments, there is perhaps a good case for its use by the small-scale breeder too – especially when it comes to breeds that may be at risk. By doing so, it is possible to maximize the number of pure-bred litters due to the fact that the practice allows a greater amount of inseminations per ejaculate of semen. When it comes to finding out more about AI, organizations such as the BPA or local pig-breeder's clubs can help you, including instructing you on the finer points of the technique itself – although for the first time or two, you may wish to ask an experienced pig-handler for assistance or even enlist the services of your veterinary surgeon.

Gestation is about 115 days or, as is often commonly quoted, 'three months, three weeks and three days'. As the big day draws closer, a sow's udder will enlarge and the teats begin to swell: indeed, at the latter stages, it might even be possible to extract some milk by the gentle use of finger and thumb; but do not attempt anything that is likely to cause her stress – it is most certainly not worth upsetting her at this stage.

LEFT Gilts must be developed enough to withstand the attention and weight of the boar before being mated for the first time.

FARROWING RAILS

Although otherwise very gregarious creatures, when a sow is due to farrow she must be given a place of her own. Even the hardiest of breeds will appreciate some overhead protection at this time and the ark or shed needs to be large enough for a pregnant sow to turn around in comfort.

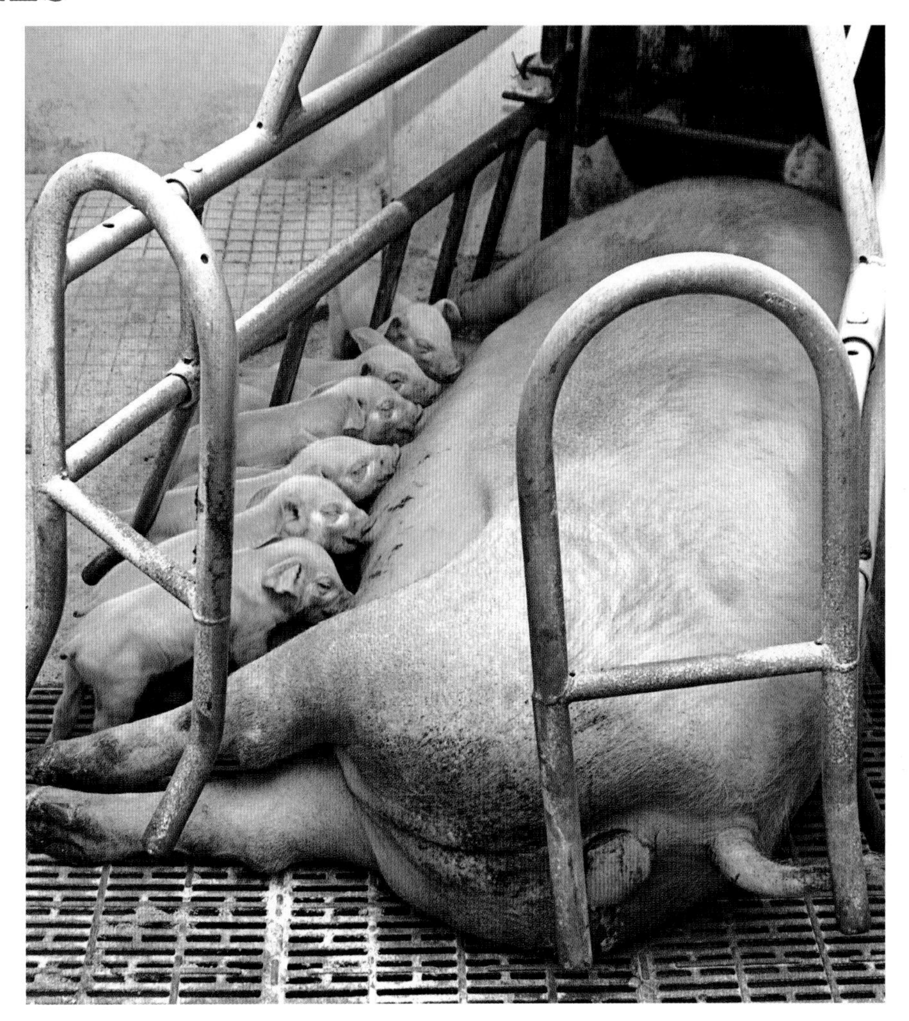

Natural instinct takes over immediately prior to giving birth and most pregnant females will use whatever material there is lying around in order to create a nest. Some experienced pig-breeders recommend that the sow should be given no floor litter at this stage, but there are probably more who say that access to material with which to make a comfortable nest is the kinder thing to do. It is however, important that not so much straw is given that the piglets can burrow under it and risk being laid upon by 'mum' as she flops down her huge body, so perhaps as a compromise, you could consider using a farrowing rail?

It is possible to buy specifically made farrowing arks that have a farrowing rail already fitted, but you could easily adapt an existing shelter. A fallowing rail fixed roughly 25cm (10in) from the wall and the same distance from the floor, will offer some protection to the piglets by preventing them from becoming squashed as their mother lies on her side. Another alternative used by some pig-breeders when farrowing indoors is to hang car tyres around the walls so that the middle of the tyre is approximately the same height as the sow when she is lying.

Farrowing crates were traditionally used in commercial enterprises where pigs were kept intensively. Because the sow's movements are restricted and she finds it virtually impossible to turn round, stand up, lie down or rest without difficulty, much of her natural instincts when giving birth are thwarted and, therefore the system is frowned upon today. There are, however, some alternatives – the Ellipsoid farrowing crate and Werribee farrowing pen, for example – that allow the sow to turn around without difficulty, lie down and, perhaps most importantly, have more contact with her piglets.

ABOVE Farrowing crates protect the piglets from being accidentally crushed but are the subject of much controversy.

POTENTIAL PIGLET PROBLEMS

It can sometimes be difficult to decide when a sow has produced the final piglet: she may have expelled the placenta, but there is just the possibility that she may have another even after doing so. At any stage too, a youngster may be born dead. Given a fair chance, it is probable that a sow will look after her offspring quite well on her own. It is, however, a sad fact that there may be other fatalities on the way, commonly as a result of the mother lying on her piglets. Fallowing rails could help prevent such misfortune, but in between, there are all manner of mishaps that might occur.

The womb temperature of a pig is 39°C (102.5°F). Imagine the stress that might occur by a litter of piglets being born during the cold winter months! In such a situation, an ambient temperature of 22°C (71.6°F) should greet their arrival. This is perhaps easiest achieved by the use of an infrared lamp. There are the options of either a bulb or emitter, but in this particular situation, the bulb is probably better – if nothing else, it gives off a warm encouraging glow, which an emitter will not. In the severest of weather conditions, it might pay to construct some sort of temporary ceiling over the heated area where the piglets naturally congregate as this will help retain the warmth given off by the lamp – bearing in mind the electrical aspect, it should of course, be made completely safe (see chapter 4).

FEEDING ISSUES

A sow will readily allow her piglets to feed and usually there are sufficient teats for each of them to do so simultaneously. Occasionally though, one or more of these teats may be 'blind' and not be producing milk. A good indication of a piglet not feeding well is if it is constantly squeaking or is listless and lethargic – piglets quite often lay claim to their own particular teat and if the rest of the litter is thriving, it may be that one poor unfortunate has found either a blind teat or one that is not producing as much milk as the rest – in which case you may need to offer that particular piglet some supplementary milk via bottle feeding. More correctly known as 'sow's milk replacer', this can be bought from most agricultural suppliers and, as it is not something you are likely to keep in stock, it is worth knowing that in an

emergency, two parts condensed milk to one part water can be used until some replacer is sourced. Goat's milk is also a good temporary substitute but cow's milk is virtually indigestible to a piglet.

Sometimes, although it is obvious that she has plenty of milk, the sow may appear reluctant to allow the piglets to drink – this might be because of their sharp teeth and it was, at one time, common practice to clip off the ends of a piglet's teeth (and also to dock their tails) at a very early age. While this practice is still permissible (except if rearing under organic rules), the American Veterinary Medical Association (AVMA) advises that teeth-clipping should only be 'performed as necessary to prevent trauma to the sow's teats and snouts of other piglets.' The British Welfare of Livestock Regulations also state that 'neither tail-docking nor tooth-clipping shall be carried out routinely but only when there is evidence ... that injuries to sows' teats ... have occurred or are likely to occur as a result of not carrying out these procedures. Where tooth-clipping appears necessary, this shall only be carried out within seven days of birth.' Any mutilations such as tail-docking will, of course, preclude a pig from being exhibited in the show ring.

Several experienced pig-breeders are of the opinion that the sow will only allow her offspring to drink so much at a time for fear of them getting colic and it is indeed possible to watch a sow lying on her side with her litter of pigs feeding quite happily before she then tucks herself up so that they can't get to her teats – this behaviour should not, therefore, be confused with her preventing access to her youngsters because of their sharp teeth.

CASTRATION

Nowadays, the castration of male piglets is not the norm, as meat animals will generally go to slaughter before testosterone becomes a problem – either in the way of a male trying to mate with its litter sisters or regarding the possibility of the meat becoming flavoured by what is commonly known as 'boar taint', a supposedly unpleasant taste and odour. Studies have, however, found that there is no reduction in meat quality when a male is left un-castrated. If you are intending, for whatever reason, to keep a male pig entire, then it should perhaps be noted that they can be more aggressive and that, when kept with females, may display sexual behaviour that could cause injuries and stress. If you do decide to castrate, consult your vet.

ABOVE A poor-doer or runt of the litter might need supplementary feeding – not an option for commercial units but a possibility for the small pig-keeper.

OPPOSITE An infrared lamp offers warmth and a focal point to very young piglets.

WEANING

In the past, farmyard piglets were always left with the sow until they were about eight weeks of age and, if you are a hobbyist with an interest in traditional breeds, you might wish to consider following similar methods. The young pigs will gain sufficient nourishment from their mother for the first two or three weeks, but they should be encouraged to eat solid food, starting with a few handfuls of milk pellets, around the age of about ten days, until, by the age of six to eight weeks, they will have effectively weaned themselves.

If you were simply to put food into the same trough as the sow she would rapidly eat it all long before her offspring had a chance to sample what might be on the menu – hence the need to create a 'creep-feed' system. What this means is allowing the piglets through gaps in a barrier that are too small for the sow to use to give them easy access to concentrates on the other side. Incidentally, when an infrared lamp has been considered necessary (see previous topic), the 'creep' area is an ideal place over which to fix it as it encourages the piglets to sleep away from their mother and thus lessen the risk of them being inadvertently laid upon.

SEPARATING PIGLETS FROM THE SOW

How you separate the young from their mother after two months is a matter of personal preference. The piglets could be removed and the sow left where she is, but it is more normal and possibly better practice to leave the little ones in their familiar pen and to remove the mother instead. Wherever possible, the sow and the newly weaned litter should be kept out

of sight and sound of each other, though this is easier said than done when you have limited space at your disposal. Once separated, the sow's milk should dry up naturally but it is important to keep an eye on her to ensure that her udders don't become enlarged or very hard; in which case, seek the assistance of your vet. She will most likely have lost condition as a result of feeding her offspring and while she will quite quickly build up weight again, keep a close eye on her to make sure that she does.

In many commercial pig units, piglets are weaned and subsequently taken away from their mother much earlier than is usual in traditional pig farming; the reason being that the sow's oestrus cycle will recommence a few days after piglets are weaned and so those with early weaned piglets can be mated sooner than late-weaning sows. The more litters per year, the greater the profit is the theory. Such a practice undoubtedly causes a great deal of stress to both the youngsters and the mother, and is not generally necessary for the smallholder or hobbyist. In addition, there can be no doubt that piglets kept with their mother for longer will build up a better immune system and will probably be generally healthier in later life as a result.

PIGLET BEHAVIOUR

Observing young piglets with their mother will also give the newcomer to pig keeping the opportunity to learn something of the natural habits of pigs. The undeniable intelligence of pigs manifests itself at a very early age and you only have to look at a litter of two- or three-day-old piglets playing together to see apparent evidence of this. Try watching a group of free-range piglets playing 'chase' and you might, for example, notice that when one of them takes a wrong turning and temporarily loses itself around the back of a partition or bale, the others will suddenly stop their game and go in search of it. And lest you think it fanciful that pigs can recognize that one of their brethren is missing, it is of course well documented that pigs can count – or at least learn tricks that make them appear to be doing so – just look at the accounts of performing pigs in the circuses and side-booths popular in the late 18th and early 19th century. One such was a pig called Toby who caused the (perhaps a little self-obsessed?) poet William Cowper to write, 'I have a competitor for fame … what is a tutor's popularity worth in a world that can suffer a pig to eclipse his brightest glories?'

ABOVE Healthy piglets enjoy playing 'chase' and will often notice if a playmate is missing!

OPPOSITE Weaned by traditional methods, piglets will suffer little or no stress.

HEALTH AND CARE

IT IS FAR MORE POSITIVE TO TALK OF HEALTH RATHER THAN OF DISEASE – AND IT IS OBVIOUSLY BETTER TO PROMOTE THE FORMER FROM THE OUTSET THAN IT IS TO TALK OVERMUCH ABOUT TREATING THE LATTER – WHICH, UNLESS YOU ARE VERY UNFORTUNATE, GENERALLY ONLY OCCURS AS A RESULT OF MISMANAGEMENT. IF YOUR PIGS ARE HEALTHY AND HAPPY IT THEREFORE FOLLOWS THAT THEY WILL BE FAR MORE RESISTANT TO BOTH MINOR AND MORE SERIOUS PROBLEMS. A GOOD DIET, PLENTY OF FRESH AIR, MENTAL STIMULATION AND A REGULAR DAILY ROUTINE WORKS EQUALLY AS WELL FOR PIGS AS IT DOES FOR HUMANS.

A HEALTHY LIFESTYLE

Given the opportunity, pigs are clean animals and it is only because of the unfair conditions in which they have been kept in the past that they have gained the reputation of being otherwise. Provided that there is sufficient space and that their houses and runs are kept clean, there should be little or no problems regarding health and hygiene. However, like riffling through an encyclopaedia of medical problems appertaining to humans, where, by the end of it, you will undoubtedly be convinced that you have every ailment known to man, a quick glance through the various Internet search engines appertaining to relevant diseases may make you wary of even the slightest sniff or cough. But, and it is a big 'but', such problems are generally found in the intensive, somewhat alien environment in which commercially produced pigs are forced to live and rarely within a small, outdoor herd.

Much is based on common sense: disinfecting the arks, for example, should be done every time new animals are moved in, but if your pigs are indigenous to your property and there is little risk of disease or parasites from outside sources, any major cleaning might need to be done only twice yearly, preferably on warm spring and autumn days. Your local agricultural supplier will be able to recommend a brand of safe, non-toxic disinfectant that contains an anti-parasitic medication. Likewise, it should be obvious that, wherever possible, pigs should be periodically moved from one paddock to another in order to avoid a build-up of parasitic worms (see chapter 4).

With a healthy lifestyle, the natural lifespan of a pig is anywhere between 8 and 15 years – a large variant admittedly, but one based on the fact that much depends on the breed and its early upbringing. Whatever, you should do all that is possible to ensure your animals have the best possible opportunities; for while as a newcomer to pig keeping, you might not yet think it possible that a pig can wheedle its way into your affections like a dog or cat, you will be surprised at how bereft the loss of one of your own can make you feel.

BELOW All pigs appreciate clean bedding.

THE DAILY ROUTINE

As with all animals, a daily routine is essential and your pigs will learn to depend on it. Once established, it should be adhered to. Most animals do not like sudden change, either in their diet, their routine, or even the person that visits and feeds them each day.

Prepare beforehand for the important morning feed because as soon as your pigs hear you they will expect breakfast – yours can wait! Much can be achieved by thinking ahead and making sure that everything you are likely to need is to hand: now is not the moment to discover you are out of pig food or that a neighbour popped round the previous afternoon and borrowed the yard broom. Having pacified them with breakfast, it's time for mucking out (in buildings with a solid floor, this should be followed by a quick hose down) and checking that all is well. Depending on the nature of your daily employment, you may be able to give a quick inspection at noon in order to see that they have plenty of water, any electric fencers are working and gates and doors remain secure.

The evening feed comes next, and afterwards plenty of opportunities for the most important part of the day – making time for your pigs. Before that though, once more with the shovel and brush, and after their sty is cleaned out, perhaps a brush down for the pigs. In today's jargon, this is called 'bonding' but it used to be considered a simple matter of good husbandry. Ensure that they have enough bedding; in warmer weather they will lie on top of a thin layer of straw, while in the colder weather they will thank you for a deep bed and will burrow out of sight in it. Finally, check them last thing before you go to bed. Your animals should be sleeping or at least preparing to, but will nevertheless anticipate your visit and perhaps another snack; nothing serious – a handful of pig nuts or dry crusts, no more. If all this seems a lot of work and bother, remember, 'there is no toil where there is love' and if you don't love them, don't keep them!

ABOVE A fork is essential for spreading fresh bedding and clearing it up again once dirty!

MOVING YOUR PIGS

There will be many occasions when you will need to move your pigs around your holding; for example, to move your free-range animals into a fresh paddock, to transport a sow to the boar, or eventually to transport your animals to their final destination.

In terms of the paddock, much can be achieved initially in the design of the layout. A single gate between each paddock will allow you, and your pigs, access from one to the other with very little problem, especially if food is involved – a few pig nuts in a bucket rattled in front of the pigs and they should follow you anywhere you wish.

MOVING BOARDS AND PASSAGEWAYS

Moving boards are a useful piece of kit – if a pig cannot see where it is going, moving it around is so much easier by their use; which is, incidentally, why many pig-keepers insist that lop-eared animals are much quieter and more biddable than those with erect or prick ears, as they cannot see what is immediately surrounding them. A moving board will help when transferring a pig from paddock to paddock, or when trying to cajole it towards the ramp of a trailer. A board is also most likely to be supplied by the organizing authorities when a pig is being exhibited at an agricultural show, and normally it will have advertising emblazoned across it! Traditionally made of plywood, a moving board is quite simple to make – just take a piece of 8mm (3/8in) ply cut in a 60cm (24in) square and keyhole a hand hole in the top. Should you want to buy one and, yes, there are such things for sale on the Internet: they are made from either exterior ply or from recycled plastic (to which can be added your farm or herd name).

Pigs that are used to being moved can be gentled along with a moving board, but if they are not so used to being handled and there are any doubts that the animals will comply, it might be necessary to construct a temporary passageway of sheets of corrugated tin, plywood or similar material tied to posts. It does not need to be too solid, but should show no gaps or the pigs will make for it with a stubborn determination the like of which will surprise you. It is worth the time and trouble to do this, as much more time, trouble and possible damage will result from an escaped animal. Once they have learned they can escape, they will always keep trying so the secret is to get it right first time! Pigs will rarely attempt to go over an obstacle, but if they see an opportunity to do so, will attempt to push

their way through or under, and an adult, with its powerful shoulders and apparent 'four-wheel-drive' capabilities, possesses a strength that you cannot possibly hope to compete with.

TRAILER TIPS

If you have a trailer to move the animals, consider putting it in the paddock for a few days with the loading ramp down and even feeding them in it – once they become used to it there will be no problems. However, for the sake of security, fit a locking tow ball or you may discover both trailer and pigs have been removed in the night. The final part of the move may well be in getting the animals to leave the trailer: once again, give them no avenue of escape, stand back and leave them to it. Being inquisitive creatures they will soon emerge, but if you attempt to drive them out, they may remember and be reluctant to re-enter the next time.

At all times make a supreme effort to keep calm and coax the animals along with soothing words of encouragement: losing your temper and, even worse, striking a pig will never work, as when hurt they can become very stubborn and worst of all, you will have done much harm to your relationship. A mutual friend told us once how he had lost his temper with a much-loved animal and it took days to restore their previous relationship. We did mention earlier how like humans pigs are!

LEFT Pigs that are used to humans and being handled are far easier to move from paddock to paddock than those that aren't.

OPPOSITE Moving boards are commonly used in the show ring, but can be useful when it comes to moving your pigs at home too.

KEEPING PIGS ENTERTAINED

The need to keep pigs entertained may at first seem a strange idea, until you have kept pigs, that is. Without wishing to be at all anthropomorphic, you should always consider the pig as being an active and boisterous child. Just like a human youngster, when a pig gets bored he gets up to mischief.

The terrible habit of tail biting in commercial herds led to the poor animal having its tail docked soon after birth. Enlightened producers now use various systems to keep the pigs entertained. One such method is ingenious in its simplicity and uses a large, heavy-duty ball made of indestructible plastic with a number of holes in the side that can be part filled with pig nuts – the pig quickly learns to push the ball about and receive a reward.

As a smallholder or hobbyist, the great advantage you have over a commercial producer is that you should be able to spend more time with your pigs and part of your daily routine should include some play with them. When cleaning out, they will love to make toys of your broom and shovel, and the hosepipe is something they cannot be forced to ignore – so why bother? Instead, make the whole process a game and much more fun for both of you. As long as your drains are efficient and the weather is not too cold, give the animals a wash and shoot the water into their mouths – they just adore it. Give them a brushing down with a stiff brush while you are about it. Having had a number of plastic balls completely destroyed by pigs, a friend hit on the idea of using a damaged bowling ball – the pigs loved it and the ball survived many generations of pigs. By using your imagination there are endless inexpensive ways to keep the animals entertained, perhaps even something as simple as a length of chain securely hung from a stout beam just about in reach of the pigs. As long as the 'toys' cannot be chewed and swallowed, you will have gone a long way towards keeping your pigs happy.

ABOVE Pigs, like children, love to play, so ensure you make time for this – and if the children want to play with the pigs, all the better!

TAKING PLEASURE FROM YOUR PIGS

Why should your pigs have all the fun with their toys and entertainment? One of the simplest pleasures pigs will afford you is in just leaning on the gate and watching them feed or play. This also the most productive time you can spend, as you will soon observe if all is well.

If there are any disagreements between members of your little herd (just like humans, they do not always get on well with their fellows) you will spot it soon enough to avoid problems later. There are always temporary disputes (usually at feeding times), but they soon disappear. Also watch how your pigs sleep: a happy group that is lying close together – sometimes even on top of one another – will have no problems, though watch if one appears out on its own and look for signs of bite marks on its ears or tail.

Pleasure also comes from the way pigs greet you – recent scientific research has affirmed that short repeated grunts are the usual way a pig says 'hello'; experienced pig-keepers could have saved them a lot of time and money as they have known it all along! There is something very satisfying about their greeting and, no matter what kind of mood you are in they will soon bring a smile back to your face. Pigs are always pleased to see you, always ready for a game and never appear to have bad days – all delightful attributes that bring pig-keepers the greatest pleasure.

Hot, tired and more than a bit cross, a pig-keeping acquaintance brought an escaped sow back to her home after a long and arduous chase. Soon his anger turned to laughter when the sow walked calmly into her sty and went and sat with her head in the corner, every bit as though she were a naughty schoolchild. He recounted this tale many times over the years and every time he convulsed with mirth. That is what pure pleasure is, a sort of innocent enjoyment that money cannot buy, the sort of pleasure you can achie regularly from pigs!

BELOW "You've had your breakfast – no away and play!"

INDICATORS OF HEALTH

A healthy pig, in addition to the points noted in chapter 3, will have a generally alert appearance and be interested in all that is going on in its immediate surroundings. The only time that it might not be active and lively is when it settles to rest in a bed of straw – in which case it will often look so relaxed that you could well be tempted to join it!

An adult pig's respiration rate is between 20–30 breaths per minute (although piglets may breathe faster than this), its heart pulsates at between 70–80 beats per minute, and its body temperature should be 39°C (102.5°F). If you think a pig is unwell, take its rectal temperature (and it may be easier said than done!), and should it be more than a couple of degrees higher than this, then seek the advice of a more experienced pig-keeper, or better still, a veterinary surgeon.

READING THE SIGNS

In a healthy pig, the coat, such as it may be, should be in good condition and the skin free of 'scurf', which may be an indication of parasites or, in paler-skinned breeds, possibly even sunburn. It is interesting to note that sometimes, and often depending on the breed, a sick pig may have a denser coat than one that is well. The exception to the rule is, of course, the Mangalitza (see chapter 2), which has a natural woolly coat, which it will most likely shed during the warmer months.

As with all animals, a pig's eyes should be bright, un-sunken, clear and without any unusual growths or discharge at their corners. The snout should be cool and moist without any undue discharge

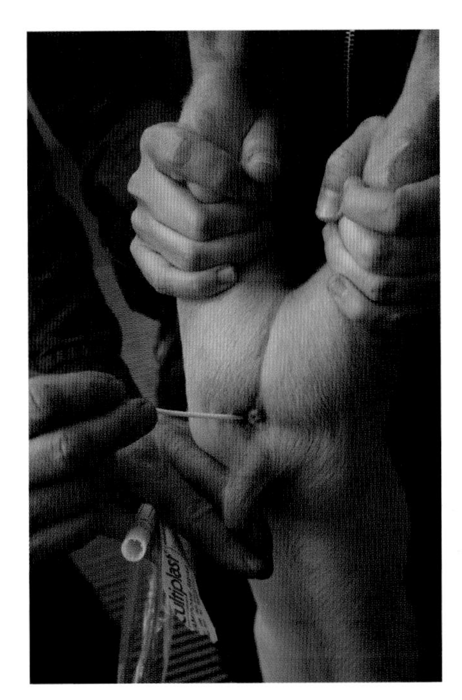

and there must obviously be no signs of coughing or wheezing. At the opposite end, it should urinate clear rather than cloudy liquid and its faeces should be solid not liquid – if not then there is potentially something wrong.

A thriving pig should never seem intent on standing hunched in a corner of its paddock or house and, being an extremely intelligent animal, it is not too fanciful to suggest that a pig that is 'under the weather' might in fact look to be suffering from the equivalent of human depression.

A lethargic pig may, of course, simply be an overweight pig, a situation that, in exactly the same way as it is with humans, should be avoided at all costs. Equally as bad as being overweight, neither should a pig look under nourished as it may be a sign of parasitic worms.

According to some experienced pig-breeders, when an adult animal's tail is straight rather than curly, then it is in need of worming. Whether there is any scientific evidence of this is another matter, but it is certainly worth mentioning as an indicator that something might be amiss (an exception might be made in the case of the Kune Kune, which has a far less curly tail than other breeds).

SEEKING HELP

If you do notice anything a little out of the ordinary, it is impor[illegible] panic. Your first approach sh[illegible] the advice of an experi[illegible] preferably by getting him[illegible] visit, althoug[illegible] provide the reassurance you need. If you are still unsure about the diagnosis or treatment of an ailment, do not hesitate to contact your veterinary surgeon. Making a friend of your vet cannot be over emphasized and it is better to make his or her acquaintance at the outset of your pig-keeping venture than it is to wait until there is a potential problem.

OPPOSITE Should medication be needed, always consult your veterinary surgeon regarding the correct product and amount to administer.

ABOVE LEFT Taking a pig's rectal temperature.

BELOW The American Veterinary Medical [illegible]ociation advises that teeth-clipping should [illegible]y be 'performed as necessary' and never as a routine procedure.

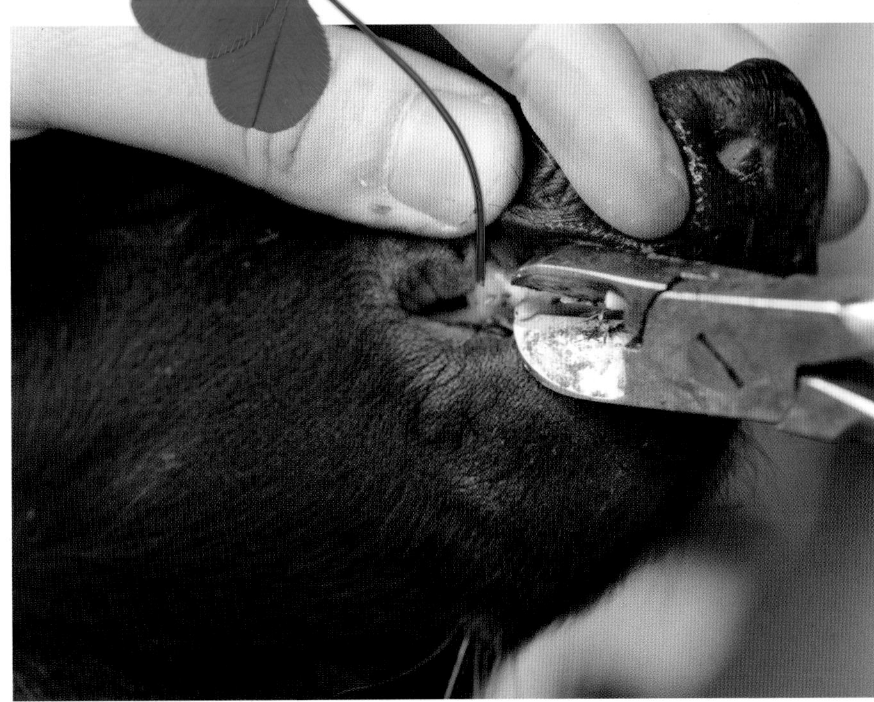

MINOR AILMENTS AND CONDITIONS

When it comes to health the outdoor pig has a great advantage; they are generally able to get sufficient exercise (which keeps their limbs strong and healthy) and occupation (so important for the mind), and are hardier by being exposed more to the elements. They are, however, arguably slightly more likely to encounter situations whereby 'nicks', cuts and parasites may occur as a result.

STRESS

The temporary equilibrium of a pig's general health can be affected by all manner of what might seem like totally inconsequential things, stress being perhaps one of the most important. Although it might seem to be a primarily human symptom of modern life, stress can in fact, cause problems both minor and major for pigs. An otherwise healthy pig can contract illness as a result of stress. The most immediate signs are a decreased appetite and possibly even a failure to drink.

Specifically, taking piglets away from their mother could, unless carefully done, result in a stress situation but, generally, until it has become well used to such things, transportation, new environments, overcrowding and administering medications can all cause a pig a certain degree of stress. Therefore, expect and look out for any obvious signs during and immediately after such operations have taken place.

PREDATORS AND PESTS

There is obviously very little that can predate on pigs, but foxes have been known to take newborn piglets from free-range or outdoor units. While a protective sow is generally more than a match for a scavenging fox, which is most likely searching for afterbirth anyway, they are great opportunists and could just about manage to take off with the runt of the litter before the mother has chance to come to its aid. An out-of-control and unruly domestic dog could also cause problems in much the same way, although the majority will be wary of a sow and an accidental encounter with an electric fence will, quite literally, result in a dog running off with its tail between its legs.

Unfortunately, where there are pigs, there are quite often rats – a far more insidious problem. Rats just love being around pigs as there is plenty of food, warmth and security offered under and around pig arks and other forms of housing. A regular poisoning routine should be your aim: obviously baiting points must be well away from anywhere that a pig has even the remotest chance of accessing. Keep a careful eye on these baiting stations as, should rats start eating the poison and then the supply dries up, they could well build up a resistance. For that reason, periodically alter the type of poison you use.

Rats are well known to be carriers of certain strains of Leptospirosis (also known as Weil's Disease) that can affect both pigs and humans – the former being commonly affected with Leptospira interrogans serovar Bratislava, which is thought can cause serious reproductive problems.

LICE

External parasites may well visit your pigs from time to time too. The signs of lice are easily visible on the skin – they are greyish-black and most commonly noticed in any folds of skin around the neck, jowl and flanks. Almost any of the commercial treatments will be effective but you will probably need to use them at least twice before completely eradicating the problem, as even the best treatments are ineffective against the eggs of lice (which can usually be seen attached to individual hairs).

While pigs always seem to like scratching and rubbing against a fencepost or corner of their shed, look out for any unusual or distressed activity, as this may be a sign of lice infestation. In pigs, lice are generally an irritation rather than a serious problem but they should, nevertheless, be [illegible]ealt with rather than left. [illegible] point of interest, the life [illegible]e of lice from the adult [illegible]ying eggs and they in turn becoming adult is around one month.

RECORD KEEPING

While any general pig-keeping notes written to yourself will be invaluable as a reference guide in future years, it is also wise to keep all receipts of animal feed purchased and records of where you obtained non-prescription-only medications (POM) if they have not been administered by a vet (agricultural suppliers, for example). It is also important to know the withdrawal periods of any substances administered to animals, especially prior to slaughter – this includes anti-fly treatment, dips and wormers. It is usually necessary to be in possession of an up-to-date veterinary medicine administration record book, which may be required for inspection by government officials. Remember that owners of pigs kept in a backyard or on a smallholding have to abide by the same legislative regulations as a commercial pig producer.

WORMS AND WORMING

Periodically moving your pigs onto new land should help to prevent a build-up of parasitic worms but some can lie dormant in the soil for months. The life cycle of such parasites usually involves a stage in the environment or in an intermediate host – for example, when it has been excreted from a pig's body via its droppings and the eggs are then eaten by snails and beetles, which in turn are eaten by pigs.

Problems are most likely to occur in sows and piglets – worms are far less likely in weaners and growers – and may include whip-worms and thread-worms. Symptoms include coughing, diarrhoea and blood in the faeces.

Wormers are readily available at your agricultural suppliers, while for those who would prefer to stay organic, there are products on the market that will efficiently solve the issue and are particularly suitable for small numbers of pigs. Leading suppliers claim that their products are herbal, non-GM, palatable and easily mixed into standard feeds as well as being gentle on the animal's digestive system. Generally, they might require feeding for seven consecutive days out of each month and a 1.5kg (3lb) jar will provide two pigs with an approximate ten-month supply at a very reasonable cost per animal.

MORE SERIOUS PROBLEMS

Unless you are absolutely certain that you have made the correct diagnosis, do not attempt any treatment yourself. While antibiotics, for example, are invaluable when used correctly, they should perhaps not be given as standard treatment, as they may merely mask the true symptoms of a disease rather than cure it. Therefore, any pig showing very obvious signs of serious ill health should be seen by a vet immediately, and any that die should undergo a post mortem in order to determine the cause of death.

PMWS AND PDNS

Postweaning Multisystemic Wasting Syndrome (PMWS) generally occurs in piglets aged between six and twelve weeks. The rate of death can be quite severe. Symptoms include weight loss, breathlessness, diarrhoea and jaundice. Unfortunately, the primary causes remain unproven, but it is interesting to note that it is a contributory factor of pigs subsequently becoming infected with Porcine Dermatitis Nephropathy Syndrome (PDNS), which affects pigs of anywhere between ten weeks and nine months of age. Symptoms for the latter include loss of weight, a dullness in the eye and skin and the appearance of crusts and scabs on the body and feet.

LAMENESS

Just like any of us on rough ground, an outdoor pig may stumble and become temporarily lame or stiff as a result. Sometimes, thankfully rarely, lameness is an indicator of something far more serious such as SVD (see opposite) or specific joint damage caused by the after effects of bacterial infections that result in problems such as Erysipelas (see opposite). Commercially produced pigs have long had a tendency towards lameness or other leg problems due to the fact that they have been bred to gain weight quickly and they are therefore unable to support their own body weight, but it is a problem unlikely to be encountered when keeping the more traditional and slower-growing breeds.

EAR INFECTIONS

Pigs sometimes suffer from haematomas in the skin of the ear. The ear will look swollen – perhaps as a result of an injury where the skin has not been damaged but nevertheless there is a build-up of blood. While it will normally dissipate naturally, the injury can sometimes become infected if the skin of the ear is then subsequently broken.

Some ear infections (canker, for example) may occur as a result of lice infestation and can generally be treated easily. Other internal ear infections are, however, potentially more serious and, if left untreated, can result in meningitis (see below).

MENINGITIS

Basically an inflammation of the meninges membranes that cover the brain, the type of meningitis to be found in pigs is usually caused by bacteria such as streptococci. It seems almost always to affect piglets, although on occasions, it may occur in adults where it is thought to be stress-induced.

Possible symptoms include a cycle of restless activity followed by periods of lethargy. There may be an unwillingness or inability to defecate or urinate, a loss of appetite, loss of balance, shivering, spasms and, typically, a continual movement of the eyes from one side to the other. Occasionally piglets might just be found dead but if veterinary treatment is prompt (usually via penicillin injections) some animals will recover.

NOTIFIABLE DISEASES

Unlikely though it is, small groups of carefully kept pigs can nevertheless, succumb to diseases that are 'notifiable' – usually as a result of outside influence. A notifiable disease is one that is considered so serious, the relevant authorities insist on knowing immediately that a suspected outbreak has occurred – in fact, it is a legal requirement. The following four are given as examples.

- **Foot and Mouth Disease (FMD)**

Unsurprisingly given its name, it causes a fever and the development of blisters in the mouth and on the feet. Animals contract the disease by either direct contact with an affected animal or contact with infected auxiliary equipment.

- **Swine Vesicular Disease (SVD)**

The clinical signs are indistinguishable from FMD in pigs – general symptoms depend to some extent on the particular strain of SVD virus and some of the milder ones can go unnoticed. Fortunately, it is thought that most of Europe is clear of the disease, and America too, where it is referred to officially as being 'a foreign animal disease'.

- **Aujeszky's Disease (also known as Pseudorabies virus)**

Not recently recorded in Great Britain, cases have, however, been notified in Northern Ireland and many other countries. It was once prevalent in the US and, while it has been eradicated in commercial operations, the virus is still found in feral pigs. A surveillance programme therefore exists to make sure that the disease is not spread from feral to domestic pigs. Far more likely in younger pigs, this herpes virus can cause problems with pregnant sows. Symptoms include difficulty in breathing, coughing, sneezing and seizures or paralysis, as both the respiratory and nervous systems are affected.

- **Swine Fever**

Caused by a virus, Swine Fever is a disease most commonly affecting young pigs, which can develop feverish symptoms, appetite loss, coughing, unusual nervousness and a discolouration of the skin – especially around the ears.

SWINE INFLUENZA

As with cases of Avian Influenza, recent problems with so-called swine 'flu caused international hyperbole in the media. In actual fact, there are several types of influenza viruses; the one most common to pigs being Type A – which is present in all countries. Although Type A can also infect other animals (including humans), the strains (sometimes known as virus sub-types) found within a single type are different and there is, as far as we are aware, no evidence that pigs spread swine 'flu to people. As it affects pigs, symptoms include a sudden onset of fever, depression, coughing, discharge from the nose or eyes, sneezing, breathing difficulties, eye redness or inflammation and appetite loss. Swine 'flu viruses are thought to be spread mostly through close contact among pigs and possibly from contaminated objects such as food buckets and the like being moved between infected and uninfected pigs.

ERYSIPELAS

This disease is infectious and occurs with varying severity – there is a mild, an acute and a chronic form. Purple blotches on the skin, often diamond-shaped, are easy to recognize but are not always present. Other signs are a high temperature and heavy, rapid breathing. At one time commonplace among pig-keepers, a better understanding of hygiene and the ability to vaccinate have all helped ensure that Erysipelas is encountered far less often than was once the case.

EVERYTHING BUT THE SQUEAK

AS QUOTED BY BUTCHERS AND PIG-FARMERS ALIKE, THERE IS 'NO WASTE FROM ANY PART OF A PIG APART FROM ITS SQUEAK!' AND THE SAYING IS TRUE BECAUSE UNLIKE MANY OTHER ANIMALS BRED FOR MEAT, IT IS REMARKABLE JUST HOW LITTLE WASTAGE THERE IS ON A CARCASS. ALTHOUGH OBVIOUSLY THE ACTUAL BREED HAS SOME BEARING, THE FINEST FLAVOURS WILL COME FROM PIGS THAT HAVE BEEN REARED IN AN OUTDOOR ENVIRONMENT AND HAVE BEEN ABLE TO FEED ON NATURALLY OCCURRING NUTS AND HERBS.

REARING YOUR OWN MEAT

Sad though it is, many animals are killed in order that we might eat. At least by rearing your own, you will know that the ones that have produced the meat for your dinner table will have had the best possible life. (It might be interesting to note that one of the authors, despite having bred and reared pigs for many years is now – admittedly for religious reasons – a vegetarian, and that the other will only buy his pork products from a local smallholder because he doesn't agree with the way commercial units rear their animals!). All things being equal though, you will, one way or another, end up with a vast array of joints, chops, bacon, sausages, cured ham, brawn (made from the head of an animal once it has been boiled, cubed and set in bowls with the 'jelly' of the boiled water) and even trotters to render down to provide the best possible natural gelatine for pork pies.

FINDING AN ABATTOIR

Taking your animals to the abattoir can never be a pleasant task. Unfortunately, local places are becoming increasingly difficult to find and the journey may involve a distance of many miles. Wherever possible, seek out two or three possible alternatives and make an appointment to visit so that you can look over their premises – although the end result is the same, you will at least be reassured that the least traumatic option has been chosen. Ask as many questions as you think necessary; for example, most experienced pig-keepers and abattoirs will recommend that a pig is starved (but never left without water) for 24 hours before killing. After choosing the abattoir best suited to you, check on the days they are geared up for pigs – it is one day in the week for some – and also to find out whether, for a little extra money, they will be prepared to cut up and process your pig to your specification. You could, of course, decide to joint your pig at home but whatever, it is important to make sure that you get your own animal back. Mix-ups do happen and there is no point in lovingly feeding and caring for your pig for many weeks only to have a strange and possibly inferior carcass

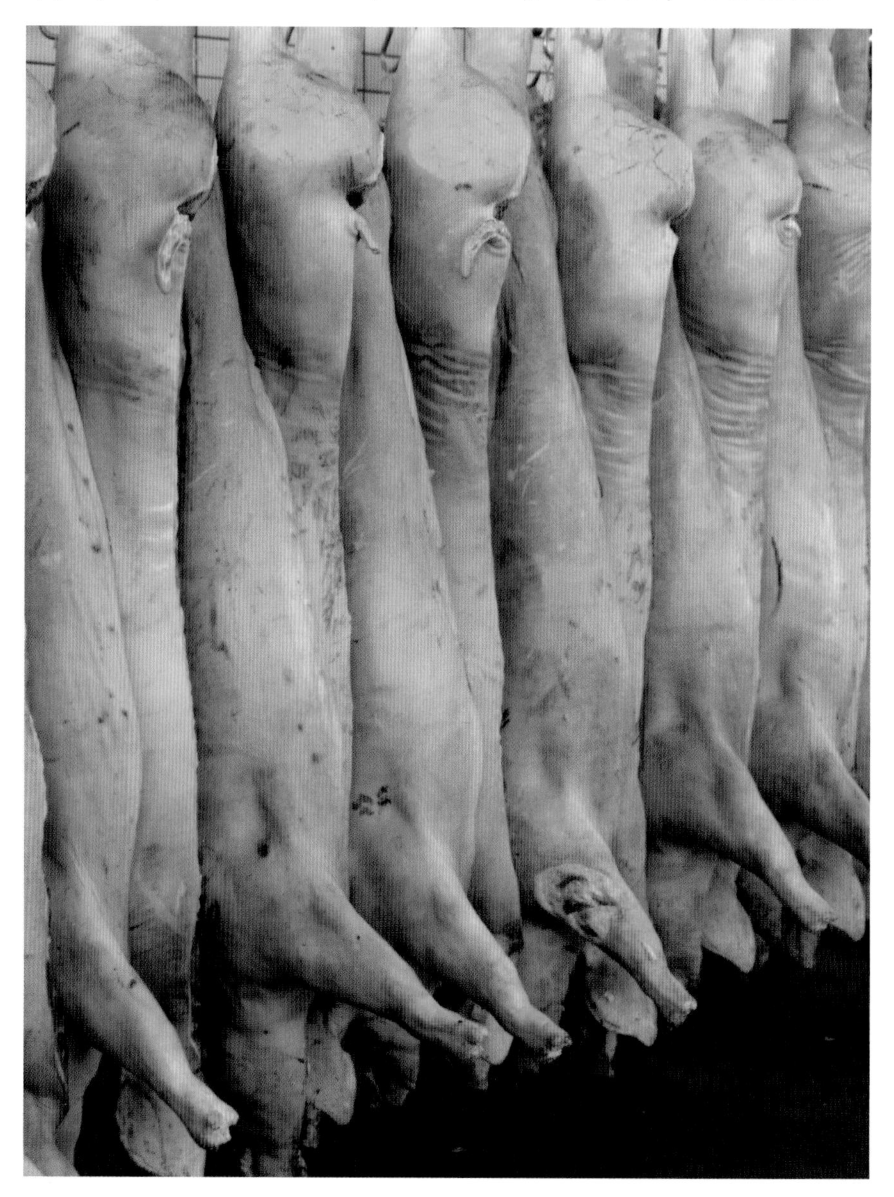

returned to you. The legally necessary slap-marks or ear-markings should, however, help prevent this.

ALLEVIATING STRESS

At one time a pig destined to feed the family would be killed at home – and, as regarding limiting stress, there was much to be said in favour of the practice. In some situations it might still be possible to do this, but legislation will no doubt stipulate that even where it is permissible, the meat should be only for home consumption and not sold on to shops, markets, friends and neighbours. Check with your local branch of the relevant government agricultural board or go online for the latest advice and regulations – which will probably also include a reminder regarding any necessary movement licences and required form-filling.

If you originally bought a couple of weaner pigs with the intention of eventually sending them to the abattoir, it would be cruel to kill one and leave the other at home by itself. Practicalities (transport, ease of butchery and dealing with the carcass) suggest that you should take both together and, if the venture has been successful in your eyes, start afresh with a new batch (once the pen and ark have been thoroughly cleaned and disinfected). Apart from anything else, sending pigs to slaughter in pairs will help with ease of loading, and the familiar companionship undoubtedly facilitates in keeping them calm on arrival. If you have your own trailer, transporting your pigs will be far less stressful if you have fully utilized the obvious opportunity to feed them in it for a few days and thereby accustom them to going up and down the ramp. Make sure your trailer is clean both externally and internally – the abattoir might object otherwise – and that there is plenty of straw to keep the occupants warm, comfortable and foot-sure.

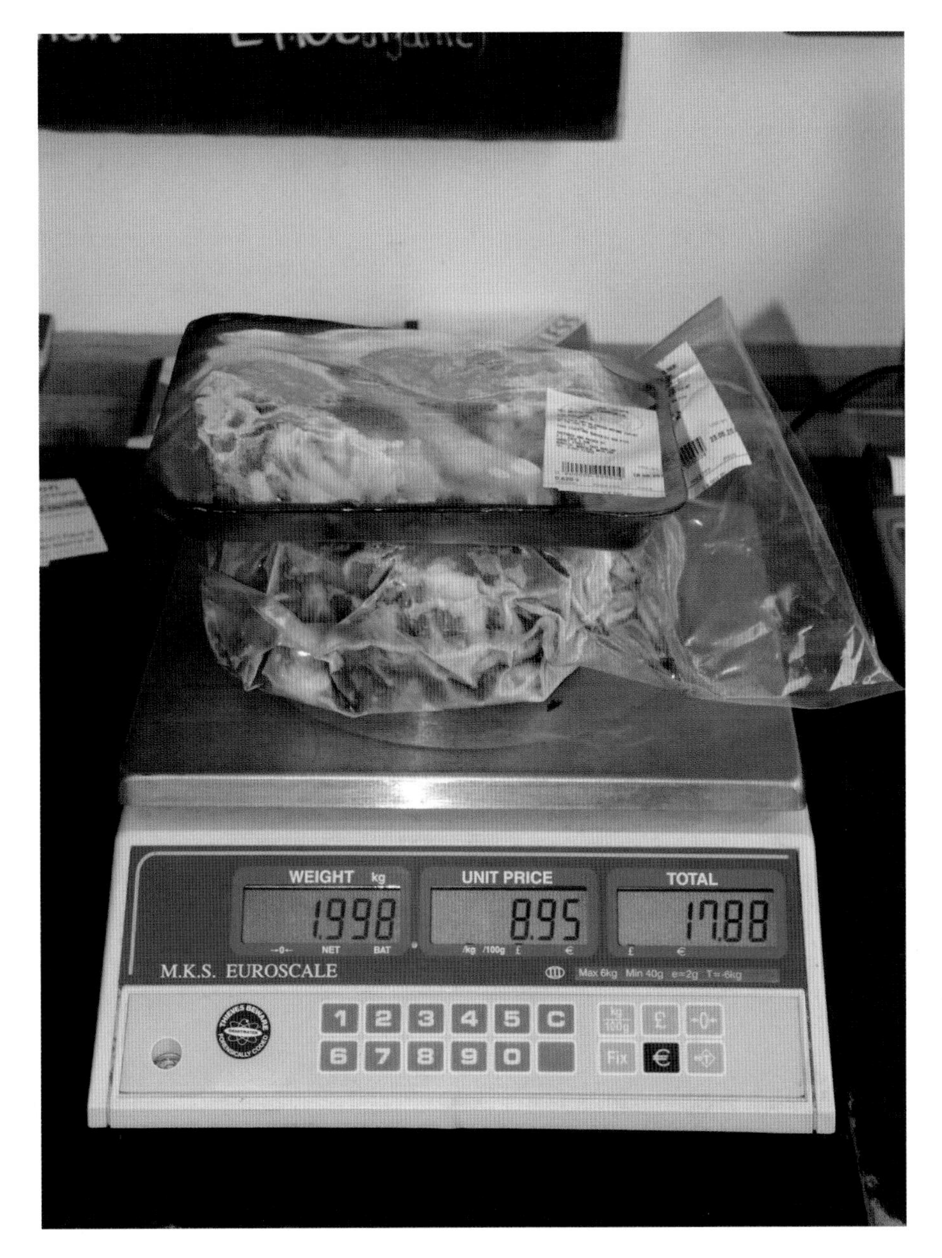

OPPOSITE Almost all parts of the pig can be used in the kitchen.

ABOVE Carefully reared pigs will always fetch a premium price.

HIGH ON THE HOG

The general parts of a pork carcass are easily identified, but actually dealing with them can be somewhat daunting for the amateur. Should you decide to go it alone rather than request the abattoir to return your pig in prepared form, it is a good idea to ask the assistance of an experienced pig-keeper or a friendly neighbourhood butcher who may, in return for some tantalizing exchange, work out and prepare such things as the shoulder, the 'hand', belly, hams and offal.

The expression 'high on the hog' almost certainly originated in the United States: it seems to have been used in a derogatory sense in order to identify those families who lived well and could afford the more expensive joints of a pig – the hams and the back. The lower parts – the knuckles, trotters and the belly pork – were deemed food for 'poor folks'. Despite such apparent segregation, the fact remains that all the parts of a pig contain the same amount of nutritious food. The belly has the same proteins as the back, the tail the same as the head – and a wise man will use every bit.

Not only are there several prime cuts on a pig, they are also very versatile; aside from the hams (of which more later), the shoulder, for example, can be used for roasting in a traditional manner or boned out to make salami. The 'hand' (the piece below the shoulder and above the trotter) may be cured, roasted or made into sausages. The loin provides bacon, smaller roasts and chops, while the belly makes joints perfect for slow

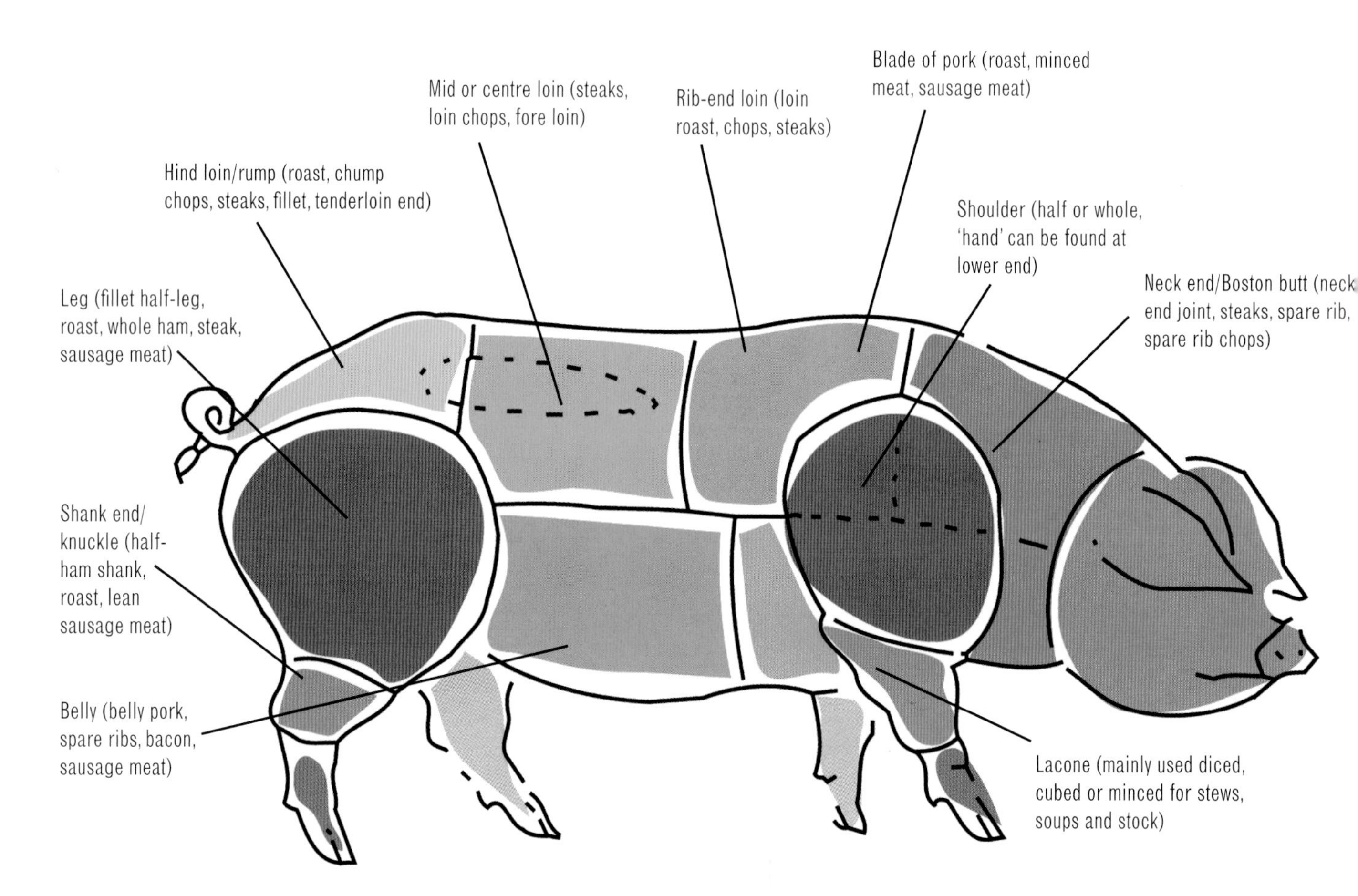

roasting, perhaps stuffed and rolled, or can be boiled and used for pâtés or potted meat. There is plenty of bacon there too – lean as well as streaky, depending from where on the belly it is cut.

You cannot think of pigs without thinking of sausages, the meat for which can come from almost any part of the carcass but it would be foolish and wasteful to use any from the best joints. A butcher will recommend which parts might prove most suitable (perhaps a mix of meat taken from the belly and a part of the boned-out shoulder) and may even be prepared to mince your sausage meat for you.

A quick read through a good cookery book will reveal many recipes for both the prime cuts, sausages and some of the lesser joints. One of the most comprehensive and easily understood books on the subject is Hugh Fearnley-Whittingstall's *The River Cottage Cookbook* (HarperCollins, 2001). In addition to giving advice on how best to prepare your carcass and some great pork recipes, it also offers very sound advice on the keeping and rearing of pigs.

ABOVE Many parts of the carcass can be transformed into sausages, and home sausage making is a delight.

OPPOSITE The main joints and points of a pork carcass.

USING LESSER CUTS

There are plenty of opportunities to use the lesser-known joints and meat 'off-cuts'. Here are just some examples:

- **Sausages** – let your imagination run wild and include herbs, spices and fruits such as apples and dried apricots.
- **Curries** – use trimmings cut and cubed as the main joints are prepared (trimmings will also be useful for homemade pork pies).
- **Burgers and faggots** – minced trimmings or sausage meat.
- **Chinese spare ribs** – yet another use for the belly meat.
- **Rillons and pâtés** – use the pork belly for rillons; a coarse pâté can be made from almost any part of the pig.
- **Salami and chorizo** – use the lean meat from the shoulders and front legs for salami; to make chorizo use a combination of the same meat and sausage mince.
- **Pork scratchings** – the rind trimmed and rolled in salt then roasted until golden and crispy.

OFFAL

While you might be turned off at the thought of it, offal is often referred to as a 'pig's fifth quarter'. The whole point of keeping a pig for meat is that almost everything is edible and while an abattoir will almost certainly return the heart, liver and kidneys, they may discard the intestines and head unless you specifically ask for them. The old adage of 'waste not, want not' was never more true than when applied to the noble pig.

When cooking or preparing any offal, make sure that it is fresh and has no apparent discolouration – it will not keep well so make sure that you have your plans in place well ahead of your pig going to the abattoir. Here are some suggested uses:

- **Blood** – black pudding.
- **Kidneys** – grilled or fried.
- **Bones and tail** – use for stock and/or soup.
- **Head** – trim, prepare and use for brawn.
- **Heart** – stuff and roast.
- **Intestines** – the ideal sausage casings.
- **Tongue** – cook and serve cold or incorporate in brawn.

LEFT Boudin Noir, Beuling, Blutwurst, Buristo, Morcilla, Blood Sausage – all are international variations of the British Black Pudding.

OPPOSITE TOP Honey-roast ham with cloves – most methods of curing include variations of molasses, vinegar, beer, herbs and spices.

OPPOSITE Depending from where it is cut, belly pork can either be quite lean or 'marbled' with fat.

CURING AND FREEZING

In most countries, there has always been the tradition that even the poorest of rural families would keep a pig and slaughter it in the autumn in order to provide food for the winter months. Cured hams and sides of bacon were hung in the rafters where the rising wood smoke from the fire helped to preserve them and also added a great deal to their flavour.

Some hams are intended to be eaten raw, others cooked. Yet more are what might best be termed dual-purpose. There are many different cures, most of which involve using brines of different strengths and for varying periods. Other variations include molasses, vinegar, beer, herbs and spices – the flavours of which may be enhanced by being air-dried or smoked over heat onto which wood chips from trees such as oak, beech, hickory and apple have been added, or herbs such as sage and bay.

The length of time a ham takes to reach maturity once cured can vary from just a few months up to two years. If left untouched and hanging in stockinette or a calico 'shroud' in a dry place anywhere between 0°C (32°F) and 15°C (60°F), hams should last for ages. Once a ham has been sliced into, keep the open end from drying out by covering it with a thick piece of greaseproof paper and re-tying the stockinette.

ADVICE ON FREEZING

Fortunately, the advent of the freezer has taken away many of the problems associated with storing meat, but even so, some care and preparation needs to be taken. If the abattoir has prepared the joints, then all should be well. Otherwise, make sure that everything is wrapped in cling-film and then in a plastic bag from which any surplus air has been excluded (by sucking it out with a drinking straw). Don't forget to pack sausages, chops, bacon slices and the like in small, easily eaten quantities.

It is important to have an organized freezer where everything is instantly available. An untidy heap of badly labelled and poorly stored joints can result in those at the bottom never getting used and eventually being thrown away. Check the freezer's interior panel for the optimum time for keeping joints and cuts of pork – and then lessen them by at least a month! It is a good idea to keep a record of what goes in and what comes out, preferably on a wipe-clean board close to the freezer.

Finally, remember that a full freezer operates far more efficiently than one that is only partially full. With careful stacking, the space between the trays and their contents is much reduced. If necessary, fill up any gaps with crumpled newspaper.

GLOSSARY

Anthelmintic A group of medications suitable for treating pigs with internal parasitic worms.

AI Commonly used abbreviation for artificial insemination.

Barrows Young, immature castrated males.

Bacon pig/baconer A grower which, when slaughtered, will dress out at anywhere between 65kg (140lb) and 80kg (180lb). Sometimes also used to refer to a long-backed pig while alive, but which is intended for this particular market.

Boar An adult male that has not been castrated and generally used for breeding.

Bowser A water container on wheels with a tow bar.

Brimming When a female is ready to be mated – *see also* Oestrus.

Carrier An outwardly healthy pig that can, nevertheless, pass on disease to other pigs in the herd.

Castration Removal of a pig's testicles, usually by surgery, whereby the testes are removed from the scrotal sac. Few other common methods of castration such as rubber bands or 'bloodless' castrators can be used on pigs due to the shape and position of the testes.

Chitlings Stomach and intestines of a pig.

'Clays' Sometimes used to describe the horny foot part of a pig.

Conformation The overall shape and stance of a pig; the standards for all pure-breeds can be found on enquiry to the relevant breed club.

Creep A portioned-off area to which piglets have access but their mothers do not.

Coccidiosis A parasitic protozoal infestation, most usually noticed in piglets.

Culling Removing and killing a poor or aged pig from the herd table.

Dressing percentage The percentage of a carcass left after it has been killed and prepared. In a pig this is quite high (*see* chapter 8).

Ear notching A method of pig identification used for pedigree pigs.

Farrowing Giving birth to piglets.

Finisher A grower over 70kg (150lb) live weight.

Flushing Some breeders increase feed to the sow in order to stimulate ovulation – not recommended to the small-scale or hobbyist pig-keeper.

Free range Access to the outdoors, sometimes restricted, sometimes unrestricted.

Gambrel Twin hook used to hoist up a pig carcass before halving and butchering.

Gestation Period of pregnancy – 115 days, traditionally given as three months, three weeks and three days.

Gilt Young female up to the stage of her first litter. When mated she is sometimes known as an 'in pig' gilt.

Grower A pig between weaning and being ready to enter a herd, sold, or being killed for the freezer.

Hand The front leg (up to the bottom of the shoulder blade), cured on the bone.

Hand-mating A term to describe taking a single sow to a boar – rather than having a boar running with a group of sows.

Hams Technically, only the back legs are entitled to be called hams.

Heat; come on heat *See* Oestrus.

Herd A group of pigs.

Herd book Usually maintained by a central organization in order to ensure the pedigrees and lineage of pig breeds and strains. Registration in a herd book is mandatory should you wish to breed pedigree pigs, enter them at shows or sell on in the future.

Hog Alternative generic name for pigs, which is quite often used in the US.

Immunity Resistance to a particular disease.

Inbreeding Used to fix desirable genes, although it is not recommended for the novice as undesirable genes may also be passed on.

'In pig' A mated female.

Lactating A sow that is producing milk for her piglets.

Line breeding Using two unrelated pigs, the offspring of which are subsequently mated together in a systematic way to produce two distinct lines of pigs based on each of the parents. The lineage of any offspring, no matter what generation, are traceable to one or other of the original parents and therefore, in theory, the gene pool of the first generation can be re-created endlessly.

Litter A group of newly born piglets. The term is also used to describe bedding and floor covering.

Moving board Useful for moving pigs around – most commonly used at agricultural shows. You can make your own from a piece of 8mm (3/8in) ply cut in a 60cm (24in) square, with a hand hole cut out at the top.

Oestrus A period of heightened sexual arousal and activity, when the sow's hormones make it ready for a successful mating.

Outcrossing When a totally unrelated pig (maybe even of a different breed) is brought in to add new blood in situations where a strain or breed has become weak through repeated line breeding or inbreeding. Like inbreeding, outcrossing can be risky, as unwanted traits may be introduced if the parentage of the new pig is unknown.

Pannage The practice of turning out domestic pigs into woodland to feed on naturally occurring foodstuffs.

Piglets Youngsters of either sex up until the age of weaning.

POM General abbreviation for prescription-only medications.

Runt The smallest, generally the weakest, piglet in a particular litter. Other traditional names include: Didler, Darling, Nisgul, Squeaker, Squabbett, Cranker, Gruntling, Dannel and Tail-ender.

Service; to serve The mating of a sow, either by natural means or artificial insemination.

Shedding The natural moulting process whereby some pigs, such as the Kune Kune, changes its 'coat' depending on the season – the emerging hair pushes out the old one.

Shoats Adolescent, newly weaned males. A term perhaps more commonly used in the US.

Slap stick A broad flat piece of wood held in the hand (often in conjunction with a moving board) to aid the gentle movement of pigs from place to place. They must not have any sharp points or protrusions.

Slip Used in some English-speaking countries to describe a growing pig between the ages of 10–14 weeks.

Snout The pig's nose and main sensory organ.

Sounders Correctly applied only to male pigs allowed to roam in groups, but often used to describe any small group of pigs (or wild boar).

Sow An adult breeding female that has given birth to at least one litter.

Stag Apparently, a male pig that has been castrated later in life.

Standing; to stand When a sow is ready to be mated.

Strain A herd of pigs carefully bred over several generations by an individual breeder.

Suckling pig A young piglet slaughtered for meat; this is often described in old books (particularly those describing Medieval and even Victorian recipes), however the practice is far less common nowadays and pigs are not generally slaughtered until later.

Tagging Registered pigs must generally have identification tags fixed to their ears. They can be plastic or metal and will most likely carry a herd mark on one side and the pig's individual number on the other.

Tusks Side teeth of a male pig that lengthen with age. Such boars are sometimes known as 'tuskers'.

Wallow Either a place made in a wet area by the pigs themselves or one created by their owner in order that they might have a mud bath – which all pigs love!

Weaners Piglets that are still with their mother but are being encouraged to feed on concentrates and become independent of her.

Weaning The process of getting piglets off their mother's milk and onto solid food before removing them completely. *See* Weaners.

Wiltshire cure A universally known method of preparing a flavoured ham – usually with salt, beer, black treacle, juniper berries and peppercorns.

PICTURE CREDITS

Alamy pp11 (The Art Gallery Collection), 30/31 (Interfoto), 34 (Jeff Morgan 12), 35 (Agripicture Images), 47 (tbkmedia.de), 52 (Caro), 67 (Trinity Mirror/Mirrorpix), 130/131 (Greg Taylor); American Livestock Breeds Conservancy p60; Animal Arks - www.animalarks.co.uk pp80; 82b; 82c; 82t; Frank Lane Picture Agency pp2 (Paul Sawer), 4/5 (Ruth Downing), 8/9 (Sarah Rowland), 10 (Ariadne Van Zandbergen), 14 (Nigel Cattlin), 15 (John Eveson), 16 (Imagebroker/Helmut Meyer), 18 (ImageBroker), 20 (Jurgen & Christine Sohns), 21 (John Eveson), 22 (John Eveson), 23 (Neil Bowman), 25 (Sarah Rowland), 27 (Primrose Peacock/Holt), 29 (R P Lawrence), 33 (Mike Lane), 37 (Bill Broadhurst), 38 (John Eveson), 39 (Sarah Rowland), 41 (Gordon Roberts), 42 (Wayne Hutchinson), 43 (Richard Becker), 44 (John Eveson), 45 (ImageBroker), 49 (Sarah Rowland), 51 (Sarah Rowland), 53 (Sarah Rowland), 55 (David T. Grewcock), 57 (ImageBroker), 58 (John Eveson), 59 (John Eveson), 61 (John Eveson), 63 (John Eveson), 64/65 (John Eveson), 72 (Richard Anthony), 74/75 (John Eveson), 77 (Jurgen & Christine Sohns), 78/79 (Nigel Cattlin), 83 (David Hosking), 85t (David Hosking), 86 (David Hosking), 87 (David Hosking), 88 (John Eveson), 90 (John Eveson), 95 (Wayne Hutchinson), 96/97 (John Eveson), 98 (Wayne Hutchinson), 101 (ImageBroker), 102 (Gary K Smith), 103 (Wayne Hutchinson), 108 (John Eveson), 110 (Ray Bird), 113 (Wayne Hutchinson), 116/117 (John Eveson), 119b (John Eveson), 120 (John Eveson), 121 (Wayne Hutchinson), 123 (Nigel Cattlin), 124 (Angela Hampton), 125b (Wayne Hutchinson), 125t (Rolf Nathaus), 133 (David Hosking); Getty Images p50 (Mark H. Anbinder/Flickr); iStockphoto pp7 (Rhoberazzi), 12/13 (Anthony Gaudio), 19 (Pavel Mozzhukhin), 24 (Ruben Hidalgo), 56 (moniaphoto), 62 (Linda Steward), 68 (Kay Ransom), 69 (Anthony Gaudio), 70 (EbertStudios), 71 (Sasha Radosavljevic), 76 (Jessica Morelli), 81bl (Jill Fromer), 81br (Lee Rogers), 81t (Jill Fromer), 84b (Don Nichols), 85b (Ilya Sevrugin), 89t (TRanger), 96 (John Bloor), 99 (Thierry Maffeis), 106/107 (tadija), 112 (Rachel Dewis), 118 (a-man17), 119t (SashaFoxWalters), 132 (Susanne Bacher), 136 (Robyn Mackenzie), 140 (Barbara Tripp), 144 (Anthony Gaudio); Mary Evans Picture Library p66; Rex Features p28b (Geoffrey Robinson); Shutterstock pp28t (Ulrich Mueller), 79 (Vaide Seskauskiene), 84t (robophobic), 89b (Steve Mann), 91 (Andre Blais), 92/93 (Popkov), 94 (Matt Antonino), 100 (Igorsky), 104 (Morgan Lane Photography), 105 (Dmitry Kalinovsky), 109 (Darryl Montreuil), 111 (dyoma), 114 (Phant), 115 (Petr Marek), 122 (Alric Bolt), 134 (Zlatko Guzmic), 135 (DBtale), 137b (sbarabu), 137t (Monkey Business Images); Thinkstock p17 (Tammy Bryngelson/iStockphoto)

ACKNOWLEDGMENTS

With grateful thanks to David C Bland of SPR Centre, West Sussex and to Stuart and Julie Barker.

ABOUT THE AUTHORS

JEREMY HOBSON

Jeremy Hobson lives in France and is a freelance author and writer specializing mainly in country matters. His numerous articles have, over the years, appeared in most of the UK's country-orientated magazines and he still writes regular columns for four of them. Since 1987, he has written over 20 books, several of which have been for David & Charles – including the highly successful *Keeping Chickens*, co-authored with Celia Lewis and *Curious Country Customs*. Jeremy has been fascinated by pigs since an early age, as his father kept Saddlebacks at their family home.

PHIL RANT

Phil Rant began his career in agriculture on a mixed farm in Berkshire before moving to the post of pigman at a bacon-rearing enterprise in Bedfordshire. After a period there he joined the Civil Service during which time he started a smallholding as a hobby where among other ventures, mainly horticultural, he kept and bred pigs for many years. On retirement from the Civil Service he created a successful landscape gardening enterprise before eventually moving to France where he now concentrates on writing, gardening and woodworking.

INDEX

Note: page numbers in **bold** refer to information contained in captions.